Advances on Nonlinear Dynamics of Electronic Systems

WORLD SCIENTIFIC SERIES ON NONLINEAR SCIENCE

Editor: Leon O. Chua
University of California, Berkeley

WORLD SCIENTIFIC SERIES ON
NONLINEAR SCIENCE

Series Editor: Leon O. Chua

Series B   Vol. 17

Advances on Nonlinear Dynamics of Electronic Systems

edited by

Arturo Buscarino
Università degli Studi di Catania, Italy

Luigi Fortuna
Università degli Studi di Catania, Italy

Ruedi Stoop
Universität Zürich, Switzerland

World Scientific

NEW JERSEY · LONDON · SINGAPORE · BEIJING · SHANGHAI · HONG KONG · TAIPEI · CHENNAI · TOKYO

Published by

World Scientific Publishing Co. Pte. Ltd.

5 Toh Tuck Link, Singapore 596224

USA office: 27 Warren Street, Suite 401-402, Hackensack, NJ 07601

UK office: 57 Shelton Street, Covent Garden, London WC2H 9HE

British Library Cataloguing-in-Publication Data
A catalogue record for this book is available from the British Library.

World Scientific Series on Nonlinear Science Series B — Vol. 17
ADVANCES ON NONLINEAR DYNAMICS OF ELECTRONIC SYSTEMS

Copyright © 2019 by World Scientific Publishing Co. Pte. Ltd.

ISBN 978-981-120-151-6

For any available supplementary material, please visit
https://www.worldscientific.com/worldscibooks/10.1142/11313#t=suppl

Preface

The historical origin of the conference Nonlinear Dynamics of Electronic Systems dates back to the collaboration between the Technical University of Dresden and the King's College of London. The first edition of NDES has been, in fact, held in Dresden in 1993. During that first edition, 22 oral presentations were given and 53 participants from Europe, United States and Japan attended the conference.

Starting from that first edition of the conference, NDES has become a fixed annual event for people working in the area of nonlinear electronic dynamical systems, but not only. The scope of the conference, in fact, has broadened during the decades, including now topics such as nonlinear system theory, communication based on chaos, applications of chaos, nonlinear control, bifurcations, nonlinear systems modeling, synchronization, neurosciences, technology, smart materials, and complex networks.

Three continents, seventeen countries, hosted NDES during the last 25 years: Dresden 93, Krakow 94, Dublin 95, Seville 96, Moscow 97, Budapest 98, Rønne 99, Catania 2000, Delft 01, Izmir 02, Scuol 03, Vora 04, Potsdam 05, Dijon 06, Tokushima 07, Nizhniy Novgorod 08, Rapperswil 09, Dresden 10, Kolkata 11, Wolfensbuttel 12, Bari 13, Albena 14, Como 15, Reykjavik 16, Zernez 17.

It is a pleasure for us and for the University of Catania to organize the 26th NDES edition in Acireale, eighteen years after the 2000 edition hosted in the Cittadella Campus in a joint event CNNA 2000. In NDES 2000, 57 oral and poster contributions and 3 invited speeches were given. In NDES 2018, we have 51 oral contributions and 13 keynote speakers.

We planned to organize the event avoiding parallel sessions in order to strengthen the community and encourage discussions and idea dissemination. The choice of Acireale is also to suit this direction, representing a peaceful place located at the geographical center of the most interesting beauty spots of south-eastern Sicily: Etna, Taormina and Siracusa.

We strongly thank the keynote speakers who accepted our invitation at NDES2018. Our deep thanks also goes to the sponsors of the conference and in particular to the Accademia degli Zelanti e dei Dafnici and its President Dr. Giuseppe Contarino.

We hope for a strong future for this event: the tradition must be maintained so that the true (not virtual) human relationships can be encouraged for the best future of the next generations.

A. Buscarino
L. Fortuna
R. Stoop

Contents

Two-phase microfluidic flow patterns and micro-channel geometry

Maide Bucolo

DIEEI, University of Catania, Italy

1.1 Introduction

The nature produces astonishing microfluidics systems that have the impressive characteristic of control the circulation of fluids in a complex network of capillaries. For applications in biology and chemistry the construction of low-cost disposable microsystems, where different fluids can circulate in a controlled manner performing a large number of tasks in a maze of micro-channels is one of the major open issue.

In the 1990 the diversification of the market of the micro-electro-mechanical systems (MEMS) gave birth to their use for biological and chemical applications. The investigation on these systems, which employed fluid flows operating under unusual and unexplored conditions, has required the rise of the interest in the Microfluidics. The progresses made in the recent years in microfluidics systems and miniaturization have led to the realization of the Point-of-care (PoC) devices,disposable low cost diagnosis tools for bio-chemical analysis [Bucolo *et al.* (2017); Sapuppo *et al.* (2008)]. That entails the miniaturization of complex fluids handling, from the single cell to the multi-phase flow, sample manipulation and integrated detection.

In finding engineering solutions for the design of portable and disposable chips for bio-chemical analysis in networks of micro-channels is fundamental a complete understanding of the phenomena governing the fluids and particles motions at the micro-scale together with the miniaturization and integration of the technologies related to mechanics, optics, electronics, fluidics and control engineering using low-cost micro-fabrication technologies [Sapuppo *et al.* (2010, 2012); Cairone *et al.* (2016b)].

This chapter will present two case studies of continuous bi-phase microfluidic flows: the mix of two immiscible fluids (air-oil or air-water), widely common in chemical applications, and the Red Blood Cells flowing in a fluid, generally used for blood sample analysis. The attention has been focused on the characterization of the flows generated inside two microchannels with different geometries. In Fig. 1(a) the two geometries considered (SMS0104, Thinxxs) are shown: a straight channel with width $w = 320\mu m$ and length $l = 16mm$ and a serpentine with width $w = 320\mu m$ and length $l = 50mm$. The channels width greater than $100\mu m$ increases the complexity of the microfluidic process due to a weak presence of turbulence and inertial effect. All that produces the flow speed up and the enhancement of the mixing, but at the same time, a loss in the process control. The results presented are related to the changes in the flow pattern due to the different geometries, in Sec. 1.2 the flow patterns in RBCs flow are presented and in Sec. 1.3 those related to the slug flows.

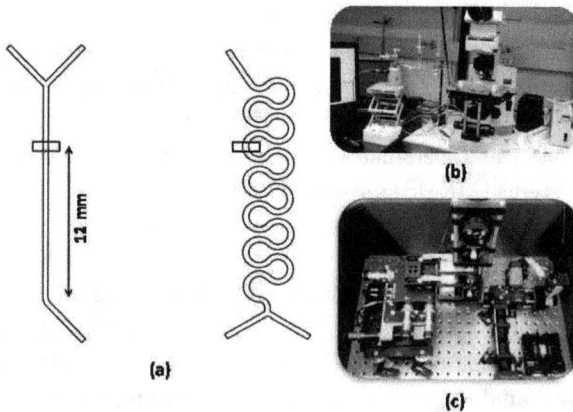

Figure 1   (a) The two geometries considered straight channel and a serpentine both with width $w = 320\mu m$. The experimental set-up is of the two case studies: (b) the RBCs flow and (c) the slugs flow.

1.2 RBCs Flows Behavioral Patterns

In both experiments, one inlet was plugged and RBCs sample was fed at the second inlet. RBCs sample, as a two-phase fluid, was obtained by diluting fresh blood taken from a hamster to a concentration of 1% (hematocrit) in a phosphate buffered saline (PBS). The hematocrit level was set at 1% to allow a clear detection of the RBCs flow by 2D imaging. A peristaltic pump (Instech P625) controlled by an ad-hoc LABVIEW interface, was used to feed the fluid-mix in the microfluidic channel. A microscope (BX51, Olympus) with a magnification of 10X coupled with a high-speed CCD (PCA 1024, Photron) was used for optical monitoring. A picture of the experimental set-up is in Fig. 1(b). Videos circumscribed an area of 1mm^2 with the micro-channel at the centre and at a distance of 8mm from the inlet. It was recorded for about 12s with a sample rate of 125Hz. Eleven experiments were carried out varying the amplitude and the frequency of the pressure strength at the inlet, a detailed description is in [Cairone *et al.* (2017a)].

A comparison of the RBCs behaviours in the two micro-channels for the same operative condition, an external oscillating pressure at a frequency $f = 0.1$Hz with an amplitude $A = 10$mmHg, is presented. That underlines the role played by the micro-channel geometry in the flow establishment.

The two movies were analysed using the Digital Particle Image Velocimetry (DPIV) and time-varying velocities vector maps showing RBCs displacements were obtained. By visual inspection of the velocities vector maps, it was possible to identify three different behavioural patterns and classified as Weak Activity, Vorticity, and Alignment [Cairone *et al.* (2018)]. Those velocities maps are then considered to investigate the RBCs flow patterns and to correlate them with the dynamics of the flow's velocity. The two signals ($\langle Vx(t) \rangle$, $\langle Vy(t) \rangle$) representative of the mean RBCs velocities, respectively on x and y directions versus time were computed. Collecting more than 100 experiments for the rectilinear geometry, a map ($|\langle Vx(t) \rangle|$, $|\langle Vy(t) \rangle|$) versus behavioural patterns was reconstructed, the details are in [Cairone *et al.* (2017a)].

Moving from one pattern to the subsequent Weak Activity, Vorticity, and Alignment it can be to notice an increase of the RBCs mean velocity and a vector re-orientation (see in Fig. 2(a) the colour and the directions of the arrows). The velocity increase can be correlated to the number of moving RBCs and their velocities. The Weak Activity was associated with a powerless and localized random stream. The vector re-orientation consisted

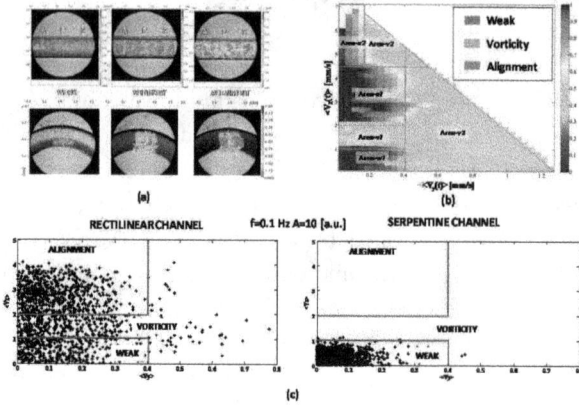

Figure 2   (a) The three RBCs behavioural patterns identified {Weak Activity, Vorticity, and Alignment} in both the straight and curved micro-channel geometries. (b) The behavioural map that establishes the relation between the RBCs flow patterns and the RBCs velocities along the horizontal ($|\langle Vx(t)\rangle|$) and vertical ($|\langle Vy(t)\rangle|$) directions. (c) The scattering plots, of the two experiments for $\{A = 10, f = 0.1Hz\}$ using the rectilinear and serpentine micro-channels.

of RBCs displacements in vortex randomly distributed along the channel length followed, the Vorticity, subsequently an ordered arrangement of their displacements along the horizontal direction, the Alignment, is reached. Starting from those distinctions, the relation between the RBCs flow patterns and the RBCs velocities along the horizontal ($|\langle Vx(t)\rangle|$) and vertical ($|\langle Vy(t)\rangle|$) directions was established by the maps shown in Fig. 2(b).

The scattering plots of the two experiments for $\{A = 10, f = 0.1Hz\}$ using the rectilinear and serpentine micro-channels, are shown in Fig. 2(c). In both scattering plot ($|\langle Vy(t)\rangle|$, $|\langle Vx(t)\rangle|$) the behavioural areas {Area-w1; Area-v1 Area-a1 and Area-v3} are highlighted.

Considering the **rectilinear case**, $|\langle Vy(t)\rangle|$ are one order smaller than those of $|\langle Vx(t)\rangle|$ but their relative variations are similar. The Alignment is strongly correlated with higher $|\langle Vx(t)\rangle|$ values because it implies a greater net flow compared with the randomly organization in the Vorticity. The behavioural areas {Area-v2 and Area-a2} were no detected for this input pressure strength.

Looking at the **serpentine channel**, as well,the dynamics $|\langle Vy(t)\rangle|$ is one order weaker than the one $|\langle Vx(t)\rangle|$. In the scattering plot both the Vorticity and the Alignment has not been reached for $A = 10$, with low input dynamics, due to the geometrical characteristics of the serpentine chip. The fluid is free to flow in a rectilinear pattern showing all the classified ways to move, alignment, vorticity and weak characteristics, whereas RBCs are slowed down, in the serpentine channel which is full of curves. The method, moreover, recognizes some whirling phenomena attributable, once again, more to the geometry of the chip than to a real vorticity. The Alignment is reached for higher input flow rate [Cairone *et al.* (2017b)].

1.3 Slug Flows Regime Formations

A continuous slug flow was generated by pumping de-ionized water and air at the Y-junction of two micro-channels. Two syringe pumps (Cetoni, Nemesys) were connected to the channel inlets, and constant flow rates were imposed.

To evaluate the effect in the slug displacement induced by the curves, the process was monitored in a position located at a distance from the Y-junction of about 4mm in the rectilinear geometry and, after nine bends in the serpentine micro-channel. A simultaneous acquisition of light intensity variations, by means of a photodiode based circuit and a standard CCD (DCU223, Thorlabs, used with frame rate of 15fps), as shown in Fig. 1(c) and detailed in [Cairone *et al.* (2016a)]. The signals used for the flow investigation are those acquired from the photodiode, giving an average information in their active area.

The signals acquired by the photodiodes range on two or three levels. The top level reveals the water presence in the channel and, the lower value the air slug passage, while some lowest peaks are for the slug's front and rear. The attention was focused on the effects on the slug displacement due to the winding geometry, other details and analysis on the experimental campaign are in [Cairone *et al.* (2016a)]. The two micro-channel geometries were considered, and the two inputs flow rates were set equal ($f = V_{air} = V_{water}$) varying in a range [0.1, 2.5]ml/min. Each experiment recording lasts 240s.

Based on the experimental setups used, three flow patterns {Slow, Fast and Annular} can be clear distinguished by visual inspection of the signal dynamics. In the Slow dynamics, the length of both air and water intervals can have significant variations during the experiment and the flow can be characterized by a sequence of smaller water intervals compared to the air intervals or vice-versa. In the Fast dynamics, the length of air and water intervals are closer than in the previous case and during the time course, the significant changes cannot be detected in their lengths but in their inter-distance. It can be interpreted as a fast train of small air/water slugs of similar lengths. In the case of Annular flows the two fluids run parallel in the microchannel and there is no slug formation. The signals related to the slow patterns can be associated to a square wave trend, and those related to the fast patterns to an oscillating behavior. It can be verified that the transition point between the two dynamics is correlated to the input flow rates and the geometries. Considering equal input flow rates, the fast dynamics were detected respectively at $f > 0.7\mathrm{ml/min}$ and $f > 1.5\mathrm{ml/min}$. Being the hydraulic resistance of the straight channel (Rh1) three times lower than the one of the serpentine (Rh2) it was expected that the transition starts before and last less in the rectilinear micro-channel compared with the other. In Fig. 3 a map that describe the changes of flow patterns lead by the increases of the input flow rate and the passage, for both geometries, from the Slow flow to the fast and finally the Annular. These passages are related with the increase of water inside the micro-channel computed by the parameter delta in the y-axis (defined in [Cairone et al. (2016a)]). That confirms the higher sensitivity to the input flow conditions in the straight channel compared with the curved one, so the robustness introduced into the process by a winding geometry.

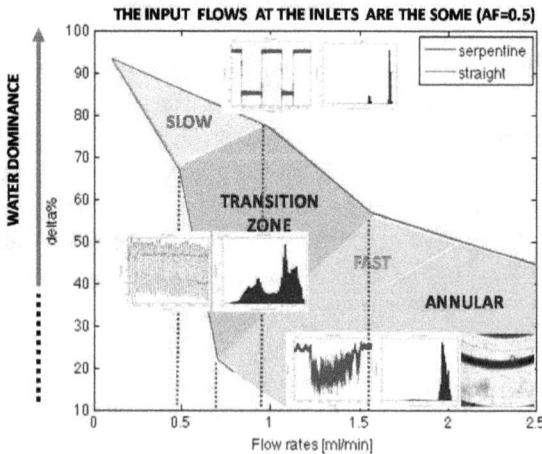

Figure 3   The map (delta vs input flow rate) describes the changes of flow patterns led by the increases of the input flow rate and the passage, for both geometries, from the slow flow to the fast and finally the annular. The parameter delta computes the increase of water inside the micro-channel.

1.4 Conclusions

In this paper a systematic experimental study on the flow patterns generation in a horizontal straight and curved micro-channels of 320μm width, where a continuous RBCs or slug flow were generated are presented. The attention has been focused on the difference in the flow displacement in a straight channel compared with a serpentine.

From all the results, a great variability in the flows behaviors and a high sensitivity to the different geometry were evident, nevertheless, the wide potentialities in the use of serpentine geometries for example to enhance the robustness to the input variations. The results have provided a characterization of the dynamics even though the nonlinearity of the process.

Unstable basis function multiplexing

Thomas L. Carroll

US Naval Research Lab, Washington, DC, USA

2.1 Introduction

Chaotic signals have both good and bad properties for communications. The great variety of possible chaotic signals has motivated many papers on chaotic communications [Kaddoum (2016)].

The problem with chaotic communication has been detection. Chaotic signals never repeat, so there are no recurrent properties to aid in coherent detection. Corron *et al.* came up with a solution to the detection problem by developing a chaotic circuit that had an analytic linear matched filter [Corron *et al.* (2010); Blakely *et al.* (2013)]. The design of the particular chaotic system allowed the chaotic output signal to be described as a linear combination of basis functions. Encoding information on the chaotic signal altered the particular linear combination. For these basis functions, an analytically derived matched filter existed, so encoded information could be recovered from the transmitted signal by passing this signal through the matched filter. The system of [Corron *et al.* (2010); Blakely *et al.* (2013)] could be produced entirely as an analog circuit, which is a potential advantage for circuits that needed to operate at high speeds, or for transmitters that needed to be lightweight and use little power.

The Corron circuit was a narrow band chaotic system. As such, it could only transmit a limited amount of information in a given time, and the narrow power spectrum was easy to detect, so it wasn't useful for low probability of detection communications. Allowing the chaotic system to run freely produced a series of basis functions that overlapped with each other, causing inter-symbol interference.

The system described here uses this concept of basis functions that have linear matched filters. This work extends the idea by using basis functions with several different fundamental frequencies, so the resulting communications signal has a broad spectrum. The basis functions are made orthogonal by a rotation, aiding in their separation in the receiver.

2.2 Linear Basis Functions

For simulations, the linear basis functions are produced by the solution of Eq. (2.1),

$$x_i(t) = e^{(-\alpha_i \beta_i t/2)} \sin(\alpha_i \omega_i t)$$
$$\omega_i = \sqrt{1 - \left(\frac{\beta_i}{2}\right)^2} \tag{2.1}$$

where $\beta < 0$ causes the system to be stable and α_i sets the oscillator frequency. Scaling the damping β by the time scale factor α_i makes all the stable oscillators have the same bandwidth. There are N total oscillators.

The linear basis functions are the time reversed versions of $x_i(t)$,

$$b_i(t) = x_i(t_{step} \times l_b - t) \tag{2.2}$$

where l_b is the length of the basis function and t_{step} is the integration time step.

2.2.1 *Orthogonal Basis Functions*

The linear basis signals $b_i(t)$ are not orthogonal, so they must be transformed into an orthogonal basis. The set of N linear basis signals of length l_b are loaded into an $l_b \times N$ matrix $\mathbf{B}_l$. $\mathbf{B}_l$ is decomposed by a singular value decomposition:

$$\mathbf{B}_l = \mathbf{U}\mathbf{S}\mathbf{V}^T \tag{2.3}$$

where $\mathbf{U}$ is an $l_b \times N$ matrix and S is a diagonal $N \times N$ matrix of singular values. The $N \times N$ matrix V will be used as a rotation matrix. The orthogonal basis functions $\chi_i(t)$ are obtained from the linear basis functions $b_i(t)$ by applying the rotation

$$\mathbf{X} = \mathbf{B}_l \mathbf{V} \tag{2.4}$$

where the individual basis functions $\chi_i(t)$ are the rows of $\mathbf{X}$.

2.3   Matched Filtering

Following from the concepts introduced in Corron *et al.* [Corron *et al.* (2010)], a linear matched filter may be designed for each of the orthogonal basis functions $\chi_i(t)$. Using the orthogonal basis means that the matched filter will have the largest possible response for the basis function to which it is matched, while the response to a different orthogonal basis function will be 0.

The transmitted signal is $s_t(t)$. The receiver receives the signal $s_r(t) = s_t(t) + \eta(t)$, where $\eta(t)$ is Gaussian white random noise. From Eq. (2.4), the i'th orthogonal basis function is

$$\chi_i(t) = [B_l \mathbf{V}]_i = \sum_{j=1}^{N} b_j(t) V_{ji}. \tag{2.5}$$

The i'th matched filter is the convolution between the time reversed i'th basis function and the received signal

$$f_i(\tau) = \int_0^{l_b} \chi_i(t - \tau) s_r(t) \, dt. \tag{2.6}$$

Combining Eqs. (2.5) and (2.6)

$$f_i(\tau) = \int_0^{l_b} \left\{ \sum_{j=1}^{N} b_j(t - \tau) V_{ji} s_r(t) \right\} dt. \tag{2.7}$$

The matrix element V_{ji} is constant, so Eq. (2.7) may be rearranged to give

$$f_i(\tau) = \sum_{j=1}^{N} \left[\int_0^{l_b} b_j(t - \tau) s_r(t) \, dt \right] V_{ji}. \tag{2.8}$$

Equation (2.8) shows that to implement a matched filter for the basis function χ_i, the signal s_r is first correlated with the linear basis functions b_i and then rotated by the rotation matrix $\mathbf{V}$.

The integral in Eq. (2.8) is the convolution between the time reversed j'th linear basis function and the received signal. This convolution is equivalent to filtering the received signal with a filter whose impulse response is the same as the j'th time reversed basis function. This filter is exactly the dynamical system of Eq. (2.1) with $\beta < 0$, or

$$\begin{aligned} \frac{dx_i}{dt} &= \alpha_i y_i \quad i = 1 \dots N \\ \frac{dy_i}{dt} &= \alpha_i \left((\beta/\alpha_i) y_i - x_i \right) + s_r(t) \end{aligned} \tag{2.9}$$

and then

$$f_i(t) = \sum_{j=1}^{N} x_j(t) V_{ji}. \qquad (2.10)$$

2.4 Antipodal Coding

For antipodal coding [Carroll (2017)], data is transmitted by multiplying each of the N orthogonal basis functions by ± 1 and summing. The orthogonality will be used at the receiver to undo the sum and recover the individual basis functions. The length of the transmitted signal $s_t(t)$ will be $K \times l_b$, where the total number of data bits to be transmitted is $N \times K$.

2.4.1 *Transmitter*

The information to be transmitted consists of a set of binary bits $\rho_i(k)$, $i = 1 \ldots N$, $k = 1 \ldots M$ where $\rho_i(k) = \pm 1$. The index i refers to the particular basis function while k indicates the particular data interval of length l_b.

To encode the binary information for data interval k, each of the orthogonal basis functions is multiplied by the corresponding binary bit $\rho_i(k)$ and summed to produce the transmitted signal $s_t(t) = s(t, k)$ $k = 1 \ldots M$:

$$s(t, k) = \sum_{i=1}^{N} \rho_i(k) \chi_i(t). \qquad (2.11)$$

2.4.2 *Receiver*

The receiver receives the signal $s_r(t) = s_t(t) + \eta(t)$, where $\eta(t)$ is Gaussian white random noise. Following Eqs. (2.9)–(2.10), the received signal drives a set of linear filters with the same value of $\beta < 0$ as in Eq. (2.1). The values of α are the same as in Eq. (2.1). At the beginning of each bit interval (length l_b time steps), the initial conditions $x_i(0)$ and $y_i(0)$ in Eq. (2.9) are set to 0. If this is not done, the signal from one bit interval will interfere with the signal from the next bit interval.

For N basis functions with a length of l_b, the $l_b \times N$ matrix $\mathbf{B}_R$ is constructed:

$$\mathbf{B}_R = \begin{bmatrix} x_1(1) & x_1(2) & \ldots & x_1(l_b) \\ x_2(1) & x_2(2) & \ldots & \\ \vdots & & & \vdots \\ x_N(1) & \ldots & & x_N(l_b) \end{bmatrix}. \qquad (2.12)$$

The matched filter outputs are found by rotating $\mathbf{B}_R$:

$$\mathbf{F} = \mathbf{B}_R \mathbf{V} \tag{2.13}$$

where the output of the i'th matched filter, $f_i(t)$, is the i'th row of $\mathbf{F}$. The estimate of the binary bit $\rho_i(k)$ may be found from $f_i(k \times l_b)$; if $f_i(k \times l_b) > 0$, $\rho_i(k) = 1$, otherwise $\rho_i(k) = 0$.

2.4.3 *Numerical Simulations*

The numerical simulations used from $N = 2$ to $N = 32$ basis functions, with the time scale factor α_i evenly spaced from 1 to 10 and an integration time step of 0.01 s. For $N = 2$, the signals from the 2 basis functions are clearly visible in the power spectrum, but for $N = 16$, the power spectrum is to be continuous. One requirement for low probability of detection (LPD) communication is a broad, featureless spectrum, which is the case for $N = 16$.

Figure 4 shows the bit error rates for the simulations when the time scale constant α was set to values from 1 to 10 (for $N = 4$, for example, α was set to 1, 4, 7 and 10). The horizontal axis is (energy per bit)/(noise power spectral density), a measure of the signal to noise ratio.

The lowest possible bit error rate for antipodal coding in additive Gaussian white noise is obtained with binary phase shift keying (BPSK), in which a single waveform is multiplied by ± 1 [Sklar (1988)].

Other forms of encoding, such as sending 1 basis function per interval, are also possible. The receiver may be built as an analog circuit, as in [Carroll (2017)], so the receiving system can be very simple.

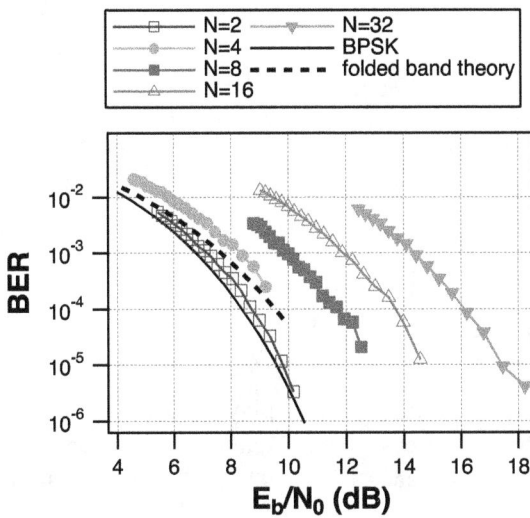

Figure 4   Simulations of bit error rates as a function of (energy per bit)/(noise power spectral density) (E_b/N_0) for detecting encoded bits. The solid line is the theoretical bit error rate for binary phase shift keying (BPSK), which indicates the optimum bit error rate for antipodal coding in additive white Gaussian noise. The time scale factor α was set to values from 1 to 10. The dashed line was the theoretical bit error rate for communication using a chaotic folded band oscillator defined in [Corron *et al.* (2010)].

Koopman operators and linear chaos

Ned J. Corron

U.S. Army AMRDEC

3.1 Introduction

Nonlinear dynamical systems can display a variety of complex behavior that is unavailable to linear systems, including coexisitng attractors, bistability, limit cycles, and chaos [Ott (1993)]. For a general theory, it is a problem that nonlinear systems can be difficult to analyze. Whereas general solutions to linear systems are known, nonlinear systems do not always admit to straightforward analysis. Thus, methods that bring linear tools to nonlinear systems are desirable. For example, the linear Frobenius-Perron operator characterizes the evolution of densities in a nonlinear system [Bollt and Santitissadeekorn (2013)]. Recently, an emphasis has been placed on Koopman operators, which also can provide a linear representation for a nonlinear dynamical system [Koopman (1931)].

In this chapter, we explore Koopman operator theory applied to a simple nonlinear dynamical system. We assume an arbitrary iterated map that may exhibit complex dynamics, including chaos. The approach yields an infinite-dimensional linear model that provides playback of a system trajectory stored as an initial condition. Surprisingly we find that this model, which is arguably trivial and unhelpful, exemplifies the mathematical phenomenon of linear chaos [Grosse-Erdmann and Manguillot (2011)].

14

3.2 Koopman Operator

Koopman operator theory considers the dynamics of observations derived from system states [Budisic *et al.* (2012)]. To illustrate, we consider a dynamical system

$$x_{n+1} = T(x_n) \tag{3.1}$$

where $x_n \in \Re$ is the state of the system at time t_n, and T is a map function. Here, we assume a scalar state for notational convenience, but the concept extends to vector and infinite-dimensional states. Furthermore, the iterated system could be derived from a flow by sampling, but it is not assumed. In general, the function T may be nonlinear and the dynamics of the iterated system (3.1) may be complex.

To investigate this system, we consider a vector of observation states

$$\mathbf{y}_n = \mathbf{f}(x_n) \tag{3.2}$$

where $\mathbf{f}$ is a vector of scalar functions. Koopman operator theory considers $\mathbf{f} \in F$, where F is a space containing all vector functions (in some sense) that are the same dimension as $\mathbf{f}$. Using equation (3.1), we then note

$$\mathbf{y}_{n+1} = \mathbf{f}(T(x_n)) = \mathbf{g}(x_n) \tag{3.3}$$

so that $\mathbf{g}$ is also a vector function with the same dimension as $\mathbf{f}$. We assume that $\mathbf{g} \in F$.

Koopman operator theory recognizes system evolution as a transformation of the observation function, $\mathbf{g} = U \circ \mathbf{f}$, where U is the Koopman operator. An important observation is that the U is linear in F. Thus, evolution of the observation states can be modeled as $\mathbf{y}_{n+1} = A\mathbf{y}_n$, where A is a linear operator. In the rare case that F is finite dimensional, the operator A is a matrix. More generally, a dynamical system that exhibits nontrivial dynamics such as chaos requires observations in an infinite-dimensional function space to completely represent its dynamics [Brunton *et al.* (2017)].

3.3 Linear Representation of Complex Dynamics

We apply a Koopman approach to the dynamical system in equation (3.1). We admit this application is not typical for a Koopman operator, which is a powerful tool that can illuminate complex problems [Budisic *et al.* (2012)]. However, our application yields an interesting and surprising result, so we proceed despite custom. We consider the observation vector $\mathbf{y}_n$ comprising the infinite sequence of scalar functions

$$y_{k,n} = T^{(k-1)}(x_n) \tag{3.4}$$

where the $k \geq 1$ is the sequence index, $T^{(k)}$ denotes k nested composi-
tions of the map function T, and we define $T^{(0)}(x_n) = x_n$. We note that
these observation functions are simply the current and future iterates of
x_n. Formally, the kth observation state evolves as

$$y_{k,n+1} = y_{k+1,n} \tag{3.5}$$

meaning each observation state simply transitions to the next iterate. Although the observations are infinite dimensional, we can still visualize equation (3.5) in matrix form as

$$\begin{bmatrix} y_1 \\ y_2 \\ y_3 \\ \vdots \end{bmatrix}_{n+1} = \begin{bmatrix} 0 & 1 & 0 & 0 & \\ 0 & 0 & 1 & 0 & \cdots \\ 0 & 0 & 0 & 1 & \\ \vdots & & & & \ddots \end{bmatrix} \begin{bmatrix} y_1 \\ y_2 \\ y_3 \\ \vdots \end{bmatrix}_{n} \tag{3.6}$$

where ones in the superdiagonal reveal the shift dynamic.

We consider the operation of the linear system for modeling the iterated
map (3.1) subject to an initial condition x_0. Using equation (3.4), the corresponding initial condition for the linear system is $y_{k,0} = T^{(k-1)}(x_0)$ for
all $k \geq 1$; thus, the initial condition contains x_0 and all its future iterates.
Equation (3.5) reveals that the linear system exhibits a shift dynamic, in
which the sequence of observation states shifts one position with each iter-
ation. Applying n iterations then yields $y_{1,n} = y_{n+1,0}$, so that the linear
system simply plays back the infinite sequence of initial conditions, much
like a tape player replays a signal previously stored on a tape.

In a practical sense, this linear representation of an iterated map is
not helpful. Using the linear system to model the iterated map requires
solving the original system to specify its initial condition. The linear model
is simply a memory storage device that plays back what has already been
computed. However, in the next section, we make a connection to the
recently identified phenomenon of linear chaos.

3.4 Linear Chaos

The system described by equation (3.5) is known and studied as a *backward shift operator* [Grosse-Erdmann and Manguillot (2011)]. This name
is explained by the equivalent representation

$$B : (y_1, y_2, y_3, y_4, \dots)^{\mathsf{T}} \rightarrow (y_2, y_3, y_4, y_5, \dots)^{\mathsf{T}} \tag{3.7}$$

meaning that its effect on an observation state vector—when transposed to
a row vector—corresponds to shifting the states backwards, or one position

to the left. On the surface, the iterated operator appears rather uninterest-
ing, since its dynamics are trivial: the infinite vector of real values march
lemming-like to the left, toward a cliff at zero, where the first element is
pushed over the edge and vanishes in an abyss. However, the iterated backward shift exhibits an interesting property that is not immediately obvious,
which is that it is a linear chaotic dynamical system.

Explicitly, the iterated backward shift meets the requirements for chaos
as defined by Devaney [Devaney (1989)]. A generalization of this definition
formally considers a metric space Y and an iterated map function $\phi : Y \to Y$. For a continuous function ϕ, the associated dynamical system is chaotic
on Y if it satisfies three requirements. First, periodic points are dense in
Y, which can be formally written as

$$\forall U \subset Y \Rightarrow \exists y \in U, n > 0 : \phi^n(y) = y \tag{3.8}$$

where $\subset$ implies an open subset. Second, the iterated function is topologically transitive, or

$$\forall U, V \subset Y \Rightarrow \exists y \in U, n > 0 : \phi^n(y) \in V. \tag{3.9}$$

Third, the iterated function exhibits sensitive dependence, or

$$\exists \delta > 0 : \{\forall y \in U \subset Y \Rightarrow \exists z \in U, n > 0 : \|\phi^n(y) - \phi^n(z)\| \geq \delta\}. \tag{3.10}$$

Of these three conditions for chaos, only the last requires a metric.

To show the iterated backward shift operator B is chaotic, we first
identify a metric space. We consider observation state vectors $\mathbf{y} = (y_1, y_2, y_3, \dots)^{\mathsf{T}}$ and $\mathbf{z} = (z_1, z_2, z_3, \dots)^{\mathsf{T}}$, where $y_k, z_k \in \Re$ for all $k \geq 1$.
We define a norm $\|\mathbf{y}\|$ using

$$\|\mathbf{y}\|^2 = \sum_{k=1}^{\infty} \left(\frac{y_k}{\mu^k}\right)^2 \tag{3.11}$$

where $\mu > 1$ is a fixed parameter and use the induced metric $d(\mathbf{y}, \mathbf{z}) = \|\mathbf{y} - \mathbf{z}\|$. We then consider the set of all bounded sequences, $Y = \{\mathbf{y} : \|\mathbf{y}\| < \infty\}$,
which completes a metric space for the backward shift operator.

To show periodic points are dense, we consider an arbitrary observation
state vector $\mathbf{y} \in U \subset Y$. We then consider a second vector

$$\tilde{\mathbf{y}} = (y_1, y_2, \dots y_N, y_1, y_2, \dots y_N, y_1, y_2, \dots y_N, \dots)^{\mathsf{T}} \tag{3.12}$$

which is constructed by endlessly repeating the first N elements of $\mathbf{y}$. For
the iterated backward shift, the initial condition $\tilde{\mathbf{y}}$ generates a periodic
orbit, with $B^{(N)}\tilde{\mathbf{y}} = \tilde{\mathbf{y}}$. Furthermore, $\lim_{N \to \infty} \tilde{\mathbf{y}} = \mathbf{y}$, so we are assured
that $\tilde{\mathbf{y}} \in U$ for sufficiently large N. Thus periodic points are dense in Y.

To show topological transitivity, we consider two arbitrary observation states $\mathbf{y} \in U \subset Y$ and $\mathbf{z} \in V \subset Y$. We then consider

$$\tilde{\mathbf{y}} = (y_1, y_2, \ldots y_N, z_1, z_2, z_3, \ldots)^{\mathsf{T}} \tag{3.13}$$

which is constructed by appending the entire sequence of $\mathbf{z}$ to the first N elements of $\mathbf{y}$. We again have $\lim_{N \to \infty} \tilde{\mathbf{y}} = \mathbf{y}$, so that $\tilde{\mathbf{y}} \in U$ for sufficiently large N. Also, $B^{(N)}\tilde{\mathbf{y}} = \mathbf{z} \in V$. Thus, the iterated backward shift is topologically transitive on Y.

Rigorous mathematical results that have shown that dense periodicity and transitivity generally imply sensitive dependence [Banks *et al.* (1992)]. However, it is still insightful to explicitly show sensitivity and a positive Lyapunov exponent. To this end, we consider $\mathbf{y}, \mathbf{z} \in U \subset Y$ such that

$$\sum_{k=1}^{N} \left(\frac{y_k - z_k}{\mu^k} \right)^2 = 0, \quad \sum_{k=N+1}^{\infty} \left(\frac{y_k - z_k}{\mu^k} \right)^2 = \epsilon^2 > 0 \tag{3.14}$$

which is possible due to the open nature of U. As such, $\|\mathbf{y} - \mathbf{z}\| = \epsilon$ and

$$\|B^{(N)}\mathbf{y} - B^{(N)}\mathbf{z}\| = \mu^N \epsilon. \tag{3.15}$$

For arbitrary $\delta > 0$, we can always choose N large enough that $\mu^N \epsilon \geq \delta$, thereby satisfying equation (3.10). Thus, the backward shift operator exhibits sensitive dependence on Y.

In equation (3.15), deviations initially grow as μ^n, where $\mu > 1$ is the norm parameter. Thus, $\lambda = \ln \mu > 0$ acts as a positive Lyapunov exponent for the backward shift. It is curious that a Lyapunov exponent is determined by the norm and is not a characteristic of the dynamical system. An understanding of this observation—and its impact on a measured entropy rate—is a goal of future research.

3.5   Conclusion

In this chapter, we apply Koopman operator theory to a general nonlinear iterated map and find its dynamics can be represented by a linear, infinite-dimensional system that acts as a storage-and-playback device. In this system, a trajectory of the nonlinear system is stored as an initial condition for the linear system. Subsequent evolution of the linear system plays back the trajectory using a backward-shift dynamic, which is shown to exemplify the phenomenon of linear chaos.

We stress that the trivial representation of complex dynamics as storage and playback is not an indictment of the merits of Koopman operator theory. Indeed, there is ample evidence that Koopman operators can provide an effective approach for finding structure in highly complex systems [Budisic *et al.* (2012)]. Instead, a conclusion of the present analysis may be that a literal Koopman approach is not well suited for describing complex behavior in simple systems, such as low-dimensional chaos, where the nonlinear nature of the system is singular and essential [Brunton *et al.* (2017)].

Mimicking railway vehicles hunting behaviors by means of sequence generators based on optimized sums of chaotic standard equations

Eugenio Costamagna[1] and Egidio Di Gialleonardo[2]

[1] *University of Pavia, Italy (retired)*
[2] *Politecnico di Milano, Italy*

4.1 Introduction

This work is devoted to investigating the possibility of mimicking sample sequences of hunting accelerations by means of a generator providing optimized sums of sample sequences derived from standard chaotic equations, as Rossler, Lorenz, Chua or similar. This kind of generators was already proposed for bursty error processes in digital transmission channels [Costamagna *et al.* (2005)], and was shown useful in video traffic simulation too [Costamagna *et al.* (2003)].

Chaotic hunting behaviors have been demonstrated and deeply analyzed using refined mechanical models, for instance in [Jensen and True (1997)], [Hoffman (2008)], [DiGialleonardo *et al.* (2014)], but it seemed interesting to perform a purely phenomenological analysis to assess the capabilities of models using statistical features derived from now from the digital transmission practice, and to develop more specific tools. In principle, similar analyses could be applied to experimental results without any knowledge of the mechanical vehicle details. Here we will use as target two acceleration sequences of 20,000 samples, one related to stable hunting behavior, and the other, at higher velocity, which we will call here lower stability sequence. Both were derived at the Politecnico di Milano from state-of-the art multi-body models. We observe first that the sequences show different behaviors both for what concern variation coefficients, long-term autocorrelations, short-term sample moving covariances and histograms, and for what concerns correlation dimensions, local variance ratio functions and Hurst parameters. Therefore, in addition to obtain good and different models for the two target sequences, it will be important to obtain similar discrimination capabilities for the models.

A summary of the work was presented in [Costamagna and Di-Gialleonardo (2018a)] and most part of the work which is perfected here has been presented in Acireale at the NDES 2018 Conference. Furthermore, some improvements and additions are made, which were at that time only planned.

4.2 Optimization of Some Important Model Features

This section is devoted to describe the results presented at NDES 2018, with some improvements obtained in the meantime, limiting ourselves to the optimization of variation coefficients, long and short-term correlation, covariances and histograms. The truly new results in the optimization of correlation dimensions and wavelet decomposition energy will be considered in the next section. Some of the new results, with the comparison of the characteristics of the stable ad of a third less stable sequence, have been presented at COMPENG 2018 [Costamagna and DiGialleonardo (2018b)]. Here, we will focus on the second, lower stability sequence, to make a complete overview and because its model was probably the most difficult to be optimized.

The traditional formulation of the variation coefficients for gaps between errors in digital transmission channels [Adoul (1974)] was adapted to the acceleration samples avoiding any reference to error probabilities saving the numerator (variances of consecutive sample blocks) but considering at the denominator only the rank of the sample blocks. This way, the original meaning is lost but the effectiveness as a means of comparison is maintained. We use up to ten Lorenz attractors with internal parameters securing the possibility of chaotic behaviors. The long term correlations are evaluated on blocks of 1500 samples, moving covariances and histograms in blocks of 5, 10, 20, 30, 40 and 50 samples. The cumulative probability distribution of the model sequences was in any case matched to the target by means of the procedure described in [Papoulis (1991)], p. 102.

Figures 5 and 6 allow comparison of target and model time histories: we see a good agreement with some imperfections, corresponding perhaps to those perceivable in some following figures.

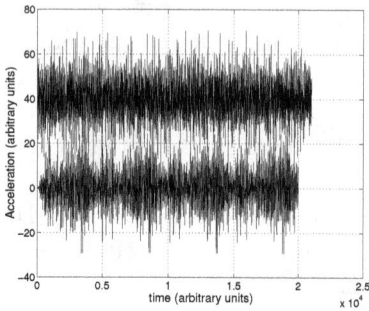

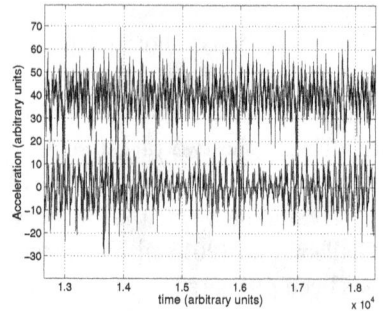

Figure 5   Time histories of the target (lower stability sequence, bottom side) and its model (biased for convenience).

Figure 6   An enlarged region of Fig. 1.

The probability distributions are shown for the target and model in Fig. 7 and the averaged moving correlations in Fig. 8: here we see a very good match in the first and similar behaviors in the second.

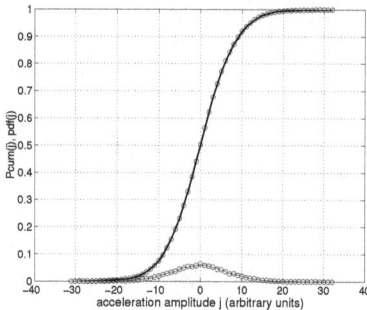

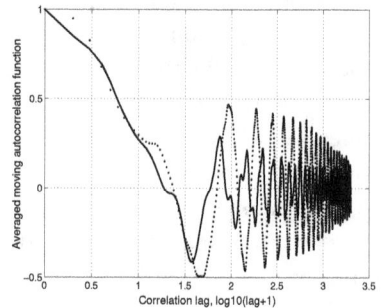

Figure 7   Probability and distribution functions for the target (dots) and model (solid line). Lower stability sequence.

Figure 8   Averaged moving long-term correlation for the target (dots) and model (solid line). Lower stability sequence.

The variation coefficients of two targets and the respective models are compared in Fig. 9. We observe a good correspondence for both the sequences, and we note that chaotic models can be of good quality and at the same time different from each other, as their targets are.

Averaged moving covariances and histograms appears in Fig. 10 and Fig. 11. Both figures show small disagreements between model and target,

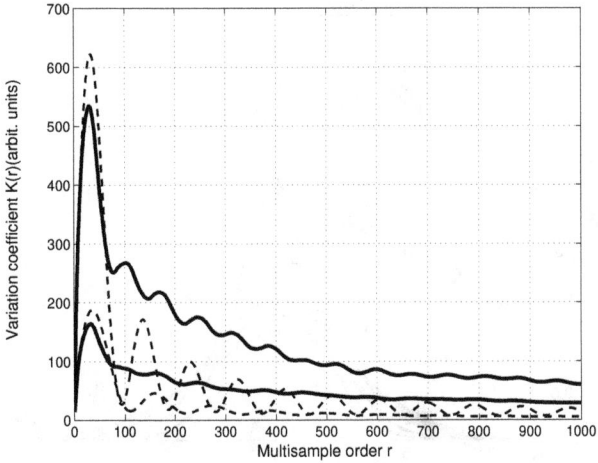

Figure 9   Variation coefficients of the stable sequence (lower curves) and of the lower stable one (higher curves). Models: thick solid lines; targets: dashed lines.

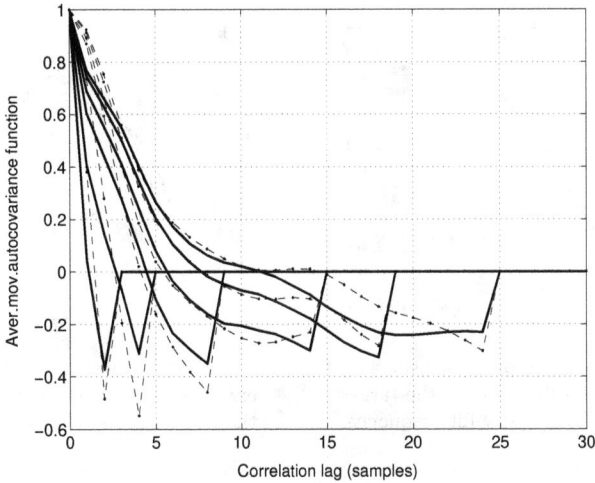

Figure 10   Averaged moving short-term covariances for the target, dots and solid lines, and model (lower stability sequence), solid lines.

the first for the largest sample blocks, the second for short and medium sizes. In fact it seems difficult to optimize at the same time variation coefficients and moving covariances and histograms. Further investigation is needed to see if this results from optimization processes trapped in local minima.

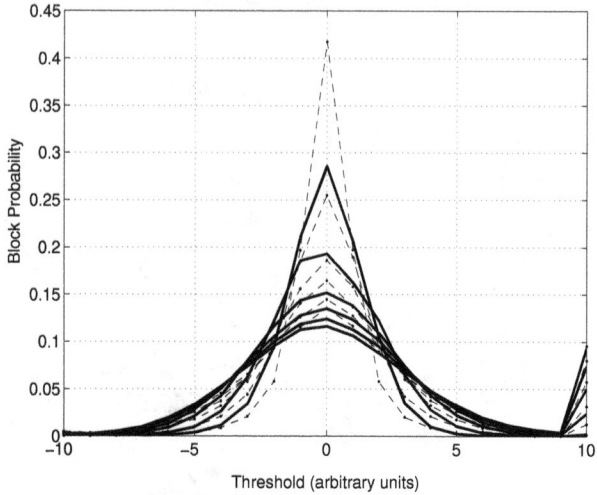

Figure 11 Averaged moving short-term histograms for the target, dots and solid lines, and model (solid line). Lower stability sequence.

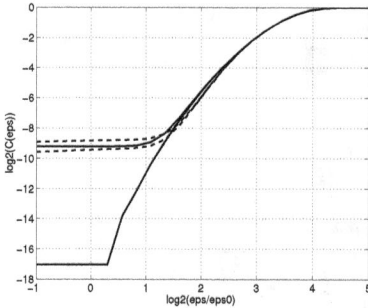

Figure 12 Correlation dimensions of the model (solid line) and of the target (dashed line). Lower stability sequence.

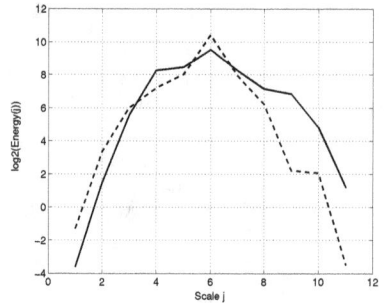

Figure 13 Wavelet decomposition energy for the model (solid lines) and for the target (dashed lines). Lower stability sequence.

4.3 Optimizing Correlation Dimensions and Wavelet Decomposition Energy

Always for the lower stability sequence, we present here the first results in truly optimizing at a same time all the previous characteristics and both the correlation dimensions [Parker and Chua (1987)] and the spectral wavelet decomposition energy [Veitch and Abry (1999)], which can be useful to explore possible long-range dependent behaviors. Although before the optimization the model curves were very far from the target, good correlation dimension results have been obtained. They are shown in the log-log plots of Fig. 12, both for the model and the target, which have been derived following [Parker and Chua (1987)] p. 1007 for 10,000 points of the sequences. The correlation dimensions values, evaluated at the branch points, are about 4.3 for both. This optimization, however, costs a reduction from about 600 to about 400 of the maximum value of the variation coefficients in Fig. 10, modest variations in the moving covariances for the larger block lengths in Fig. 11, and a modest asymmetry of moving histograms for intermediate block lengths in Fig. 12.

The optimized wavelet spectra are shown in Fig. 13. Before optimization, the largest differences were found on the right side of the figure, and then they appear significantly reduced. The curves depend significantly on the call parameters of the calculation routine, and the goodness of fit was acceptable for the model but often bad for the target; confidence intervals were always large. Paremeters alpha were in any case in the range from 2 to 2.5; fractal dimensions in the range from 3.3 to 3.7, which is compatible with the correlation dimensions; Hurst paremeters in the range from 0.3 to 0.7: data are for now too uncertain to draw any conclusion.

4.4 Conclusion

We hope to have shown some capabilities of the modeling method. Further studies have to be made, trying to improve the results and to highlight results and limits of this type of purely phenomenological models, in particular to explore the capability of mimicking unstable systems utilizing their possible chaotic behavior.

Applications of chaos theory in chaotic cities

Salih Ergün

ERGTECH

5.1 Introduction

One of the basic problems we face in everyday life is that cities become increasingly chaotic. The main reasons for chaotic cities are population growth, climate events and traffic. At this point, the most important candidate that can be used to solve the problem of chaotic urbanization is the chaos theory, known as the complex nonlinear dynamic phenomenon. This article will explain the use of chaos theory in C3PO [Mets (2014)], a EU project implemented in Istanbul, and will discuss the analysis of traffic and the recommendations for traffic within the scope of the C3PO project.

5.2 Description

Istanbul's population has left 145 countries behind with an increase of 200 thousand persons per year. (Turkey's population: 80 million - Istanbul's population: 15 million.) 20 supermarkets, 40 gyms, 20 libraries and 50 mosques are needed to accommodate these migrating people. Thanks to these accommodation units, the needs of the growing population are resolved and balanced. This problem can be solved by chaos theory, the problem of complex systems passing through a certain cycle and repeating certain situations.

Another cause of cities becoming chaotic is climate events. In July 2017, 20 minutes of heavy rainfall in Istanbul and life paralysis as a result of the storm. There were 40 kilograms of rainfall per m2, 20 thousand vehicles damaged, and 50 floods per minute [Service (2018)]. As a result, a damage bill of 1.2 billion turkish liras appeared. Whereas Lorenz, in 1963, presented a model [Lorenz (1963)] that showed changes in temperature and wind speed and could be used in weather forecasting, and discovered "strange attractor" when he found numerical solutions for this model.

The chaos theory that emerged in the 1970s has already influenced various aspects of real life in the short term (weather, traffic) and continues to influence all sciences. For example, it helps to answer previously unsolved problems in quantum mechanics and cosmology. In addition, it has revolutionized the understanding of heart arrhythmias and brain functions.

The most basic reason for cities becoming chaotic is traffic, and this problem is addressed within the scope of the C3PO project [Mets (2014)]. The number of vehicles registered on the trailer has exceeded 21 million and annual fuel consumption is over 20 million tons. In the last 10 years, about 12 people per day (42.5 thousand people in total) lost their lives in traffic accidents in the country and 555 people were injured [Institute (2014)].

5.3   Results

The chaos theory, time series analysis and bootstrap methods were used in the traffic modeling studies within the scope of C3PO Project. The historic Kyoto city in Japan is famous for its grid pattern of roads as depicted in Fig. 14. In 2013, studies by Aihara *et al.* [Suzuki *et al.* (2013)] have shown that all Kyoto City is treated as a Ising model in a two-dimensional cage. It has been shown that traffic lights have Ising-like chaotic dynamics and can be modeled chaotically. This model is predicted to form a starting point for revealing the statistical mechanics of traffic lights.

In 2013 Aihara *et al.* developed an optimal "on-off" law for traffic signals [Aihara *et al.* (2014)]. In this study, the Lyapunov function method and the dual optimization method performed well to regulate traffic congestion. Dual optimization method is the optimal solution even for large traffic networks.

Within the context of the C3PO Project, traffic analysis studies were carried out in the Istanbul Pendik urban transformation area and for this purpose two years of traffic data were collected and processed. Beside this, pedestrian and vehicle traffic analyzes were made from camera images and a system which can extract traffic density from camera images online was designed.

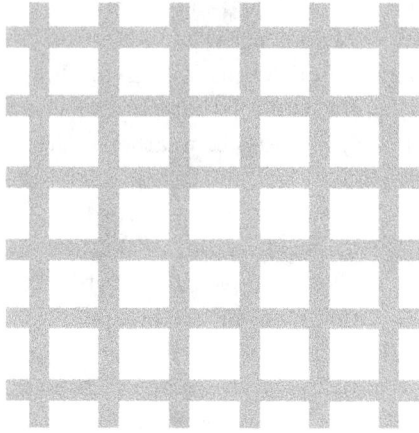

Figure 14   Grid pattern of roads in Kyoto city.

As a result of these studies, it was determined that the Pendik urban transformation area can not be modeled with Ising-like chaotic dynamics [Suzuki *et al.* (2013)] because it is not in the grid structure given in Fig. 14. It has been revealed that Lyapunov function method or dual optimization method [Aihara *et al.* (2014)] can not be used in order to arrange the traffic congestion on the side. The first concrete output of the C3PO Project is to find out that, for the use of Aihara's methods [Suzuki *et al.* (2013)]-[Aihara *et al.* (2014)], roads should be in grid structure not in a narrow area but across the all district.

The collected data and camera analysis data were processed using the bootstrap method [Zoubir and Iskander (2007)] and confidence intervals were determined for the traffic intensity on a time basis (weekday, weekend, neighborhood market, Friday, holiday etc.).

The Bootstrap method is a convenient solution that allows to determine the confidence intervals for parameters such as variances or probability distributions of parameter estimators. This useful feature makes the use of the bootstrap method ideal where numerical data is particularly short and the underlying data distribution is unknown. As a result, the upper and lower thresholds of traffic intensity have been analyzed for each street on a time-based basis.

Traffic density was analyzed at the busiest street in Pendik downtown and corresponding confidence intervals of the mean of traffic density are given in Fig. 15 for the direction of travel and the direction of rotation, respectively. As a result of the Bootstrap analysis, the direction of travel

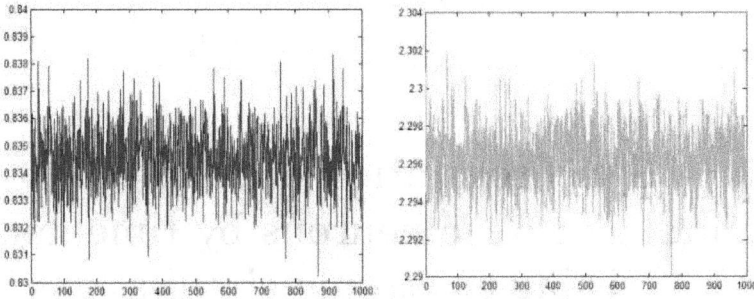

Figure 15   Confidence intervals for the mean of traffic density at the busiest street in Pendik downtown.

traffic density was 2.9 times of the direction of rotation. It is reported in the project results that, at least three lanes in the direction of travel should be built for a single lane road in the direction of rotation.

5.4   Conclusions

In conclusion it is reported that together with the urban transformation, the roads of the entire area should be transformed into the grid model and the traffic congestion can be eliminated by using the chaos theory. Competence analysis was conducted for each street as a result of the confidence intervals determined by the bootstrap method. As a result, by means of bootstrap analyzes, proposals such as adding and subtracting lanes and closing the vehicle traffic is presented in the project result report and traffic bottleneck will be eliminated on a time basis.

Synchronization of coupled Hindmarsh-Rose neurons by time-delay using electronic circuits

J. Romo-Aldana[1], J.H. García-López[1,*], R. Jaimes-Reátegui[1],
G. Huerta-Cuéllar[1], R. Chiu[1], C.E. Rivera-Orozco[1] and A.N. Pisarchik[2]

[1] *Centro Universitario de los Lagos, Universidad de Guadalajara, Mexico*
[2] *Center for Biomedical Technology, Technical University of Madrid, Spain*

6.1   Introduction

The pioneering work of Pecora and Carrol [Pecora and Carroll (1990)], where they characterized synchronization of two chaotic oscillators initiated from distinct initial conditions, aroused growing interest of many researchers in studying this topic (for comprehensive review on synchronization [Boccaletti *et al.* (2018a)] and references therein). Synchronization gained much attention in diverse fields of science and applications, such as chemical reactions [Shabunin *et al.* (2003)], electrical circuits and systems [Romo-Aldana *et al.* (2017)], secure communications [Milanović and Zaghloul (1996)], image processing [Xu *et al.* (2014)], biology [Cooper (2004)] and neuroscience [Gray *et al.* (2013); Meister *et al.* (1991)]. In particular, different neural models were developed [Hodgkin and Huxley (1952); FitzHugh (1961)] to simulate the behavior of coupled neurons depending on both the coupling strength and time delay in the coupling. Comprehensive research on synchronization between coupled neurons, based on the mathematical models were performed in the recent years in order to better understand the brain functionality [Simonov *et al.* (2013); Sausedo-Solorio and Pisarchik (2014); Gerasimova *et al.* (2015); Pisarchik *et al.* (2017); Sausedo-Solorio and Pisarchik (2017); Andreev *et al.* (2018, in press); Pisarchik *et al.* (2018)]. However, too little attention, to our knowledge, has been paid to a study of neural synchronization in the presence of a delay in the synaptic coupling.

In this chapter, we describe synchronization of two delay-coupled Hindmarsh-Rose (HR) neurons [Hindmarsh and Rose (1984)] implemented with electronic circuits. The chapter is organized as follows. First, in Sec. 6.2 we introduce the mathematical model for a single HR neuron and two delay-coupled neurons. Then, in Sec. 6.3 we describe the experimental implementation of the model in the form of electronic circuits. The main results of the synchronization study are presented in Sec. 6.4. Finally, the conclusions are given in Sec. 6.5.

6.2 Model Description

The HR mathematical model is a simplified version of the Hodgkin-Huxley model to mimic dynamics of a neuron membrane. In this work, we study synchronization of two HR neurons bidirectionally coupled with time delays, that can be modeled by the following equations:

$$\dot{x}_{1,2} = y_{1,2} - ax_{1,2}^3 + bx_{1,2}^2 - z_{1,2} - g_{2,1}(x_{1,2}(t - \tau_1) - x_{2,1}(t - \tau_2)) + I_{1,2},$$

$$\dot{y}_{1,2} = c - dx_{1,2}^2 - y_{1,2},$$

$$\dot{z}_{1,2} = r\left(s(x_{1,2} - x_0) - z_{1,2}\right),$$

$$(6.1)$$

where the variables $x_{1,2}$ represent the membrane potentials of two coupled neurons, x_0 stands for the resting potential, the variables $y_{1,2}$ represent fast currents of Na^+ and K^+, and the variables $z_{1,2}$ represent slow currents of Ca^{2+}. The parameters $I_{1,2}$ are the external currents and the parameters a, b, c, d, s, r are real constants, $g_{2,1}$ are the synaptic strengths, and τ_1 and τ_2 represent time delays in the information flow.

6.2.1 *Dynamics of a Single HR Neuron*

For better understanding of the dynamics of a single HR model, we plot in Fig. 16 the bifurcation diagram of inter-spike intervals in the time series x_1 as a function of the external current I_1. In this case, we take $g_{2,1} = 0$ and the parameters $x_0 = -1.6$, $a = 1$, $b = 3$, $c = 1$, $r = 0.006$, and $s = 4$. By changing the external current I_1 from 1 to 4, the HR model exhibits different dynamics behaviors, such as regular spiking (Fig. 16(i)), regular bursting (Fig. 16(ii)), and chaos (Fig. 16(iii)), that can be distinguished in the bifurcation diagram of inter-spike intervals.

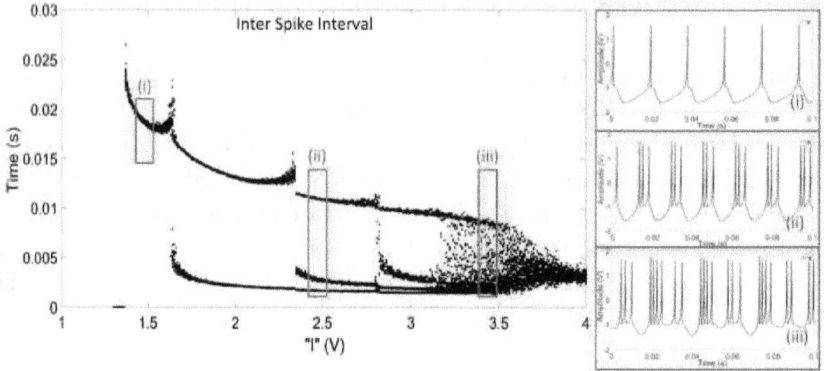

Figure 16 (Left panel) Bifurcation diagram for a single HR neuron. (Right panel) (i) Regular spiking at $I_1 = 1.5$ V, (ii) regular bursting at $I_1 = 2.4$ V, and (iii) chaotic bursting at $I_1 = 3.2$ V.

6.2.2 Two Coupled HR Neurons

In the case of two coupled HR neurons modeled by Eq. (6.1), we use the same parameter values x_0, a, b, c, d, s, r as in the case of a single neuron. For unidirectionally coupled neurons, the coupling parameters are $0 \leq g_1 \leq 1$ and $g_2 = 0$, whereas for bidirectionally coupled neurons $0 \leq g_{2,1} \leq 1$. The schematic representations of unidirectional coupling, also known as master-slave configuration, is shown in Fig. 17(a) and bidirectional coupling in Fig. 17(b).

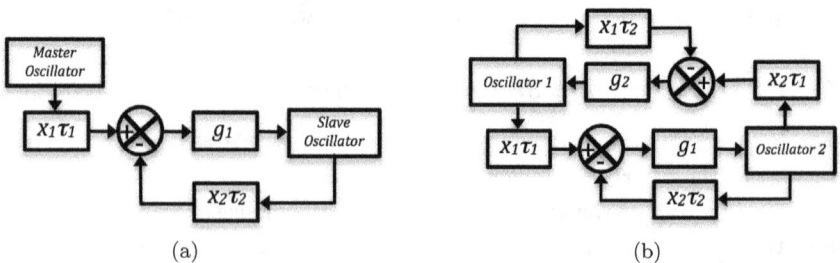

(a) (b)

Figure 17 Arrangement for coupling configurations in cases of (a) unidirectional and (b) bidirectional coupling.

6.3   Electronic Implementation of the HR Model

In this work, we implement the mathematical HR model in a physical form using operational amplifiers to construct the electronic analog circuit shown in Fig. 18. To design this circuit we use the Multisim^{TM} software [Romo-Aldana *et al.* (2017)].

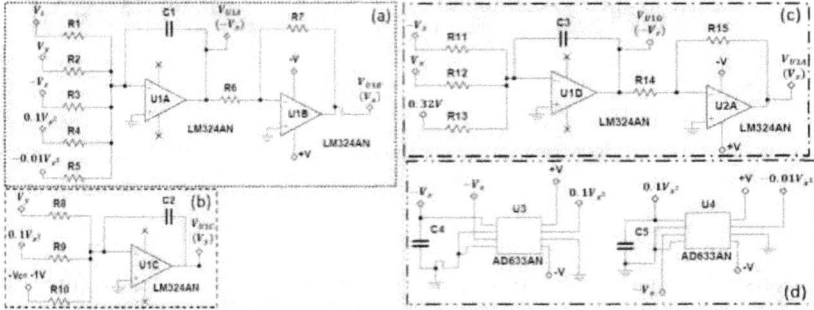

Figure 18   Implementation of HR model in the electronic circuit. The circuits simulating (a) membrane potential x, (b) fast current y, (c) slow current z, and (d) multiplier for variable x.

6.4   Results

In order to characterize the type of synchronization, we use similarity function $S^2(\tau) = \langle[x_2(t+\tau) - x_1(t)]^2\rangle/[\langle x_1^2(t)\rangle\langle x_2^2(t)\rangle]^{1/2}$ defined as the normalized averaged-in-time difference between the variables x_1 and x_2 of the two neurons, where τ is a compensation time used to compare waveforms of HR neurons with anticipation. If the variables x_1 and x_2 are coupled and $S^2(\tau) = 0$ for $\tau = 0$, then these variables synchronize. In other cases, it is necessary considering the minimum value S_{min} of the similarity function $S^2(\tau)$ at certain τ. Lag or anticipated synchronization is observed when $\tau < 0$ or $\tau > 0$, respectively.

6.4.1   *Unidirectionally Coupled HR Neurons*

In order to study synchronization between action potential x_1 and x_2 in the master-slave configuration (Fig. 17), at $I_1 = I_2 = 3.2V$ we measure the minimum similarity function S_{min} in the parameter space of $(\tau_1, g_{2,1})$ with $\tau_2 = 0$ and $g_2 = g_1 = g$. In the figure, we can see different color

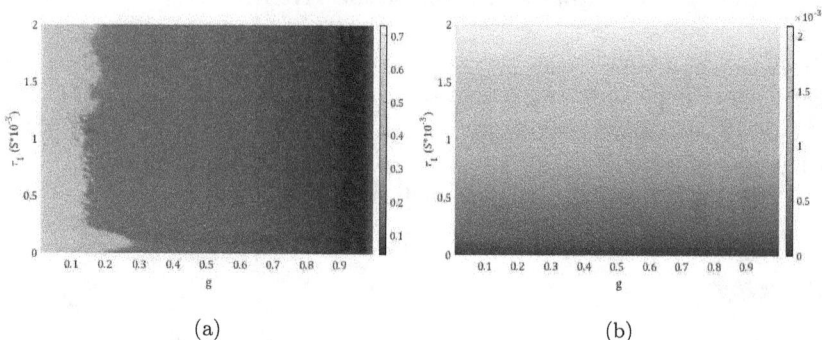

(a)   (b)

Figure 19   Analysis of synchronization for unidirectional coupling at (a) $\tau_{S_{min}}$ of x_1 vs x_2 for $I_1 = I_2 = 3.2V$ and (b) coupling strength g vs τ.

regions where each color represents the value of the similarity function S_{min} (see color bar in Fig. 19(a)). The region in blue for $S_{min} \approx 0$ means that the action potentials x_1 and x_2 are in a complete synchronization state. Whereas, the regions in yellow, green and light blue indicate that the variable x_1 and x_2 are in phase synchronization.

Next, to know whether the action potentials x_1 and x_2 are in lag or anticipated synchronization, in Fig. 19(b) we plot the compensation time $\tau = \tau_{S_{min}}$ where the similarity function reaches its minimum value S_{min} in the parameter space $(\tau_1, g_{2,1})$. The analysis of this figure is the same as Fig. 19(a), i.e., for the minimum value of $\tau < 0$ the variable x_1 and x_2 are in anticipated synchronization, whereas for $\tau > 0$ these variables are in lag synchronization, and for $\tau = 0$ complete synchronization takes place (see color bars for each case in the figure).

6.4.2   *Bidirectionally Coupled HR Neurons*

In Fig. 20(a) we plot synchronous areas in the case of bidirectional coupling for $I_1 = I_2 = 3.2V$ in the parameter space $(\tau, g_{2,1})$ with $g_2 = g_1 = g$. The analysis is the same as in Fig. 19(a). Figure 20(a) shows the minimum similarity function S_{min} with several colors. The blue color means that the variables x_1 and x_2 reach complete synchronization, while other colors indicate phase synchronization. Similar to the previous section, the interesting result is present in Fig. 20(b) where we plot the compensation time $\tau_{S_{min}}$ in the parameter space of (τ, g).

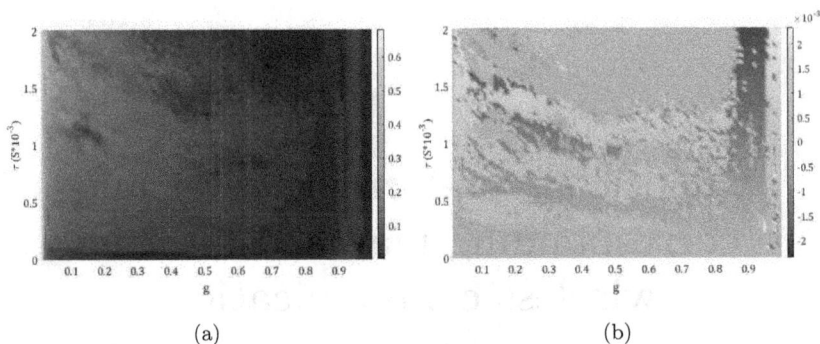

(a)   (b)

Figure 20   Analysis of synchronization for bidirectional coupling at (a) $\tau_{S_{min}}$ of x_1 vs x_2 for $I_1 = I_2 = 3.2V$ and (b) coupling strength g vs τ.

6.5   Conclusion

The time delay and coupling strength are very important parameters to control synchronization and dynamical behavior of coupled HR electronic circuits. We analyzed the dynamics a single HR neuron and two HR neurons coupled in unidirectional and bidirectional configurations with time delay. Different types of synchronization have been identified and analyzed. We have shown how the delay time and the coupling strength affect synchronization of electronic signals of two coupled HR electronic circuits.

Acknowledgments

The authors acknowledge support from the University of Guadalajara for financial support under the project R-0138/2016, Agreement RG /019/ 2016 UdeG, Mexico. A.N.P. thanks the Ministry of Economy and Competitiveness (Spain) for support through the project SAF2016-80240.

Chaotic shape-forming filter and corresponding matched filter in wireless communication

Hai-Peng Ren[1], Chao Bai[1] and Celso Grebogi[1,2]

[1] *Shaanxi Key Laboratory of Complex System Control and Intelligent Information Processing, Xian University of Technology, China*
[2] *Institute for Complex System and Mathematical Biology, SUPA, University of Aberdeen, Aberdeen, United Kingdom*

7.1 Introduction

Since chaotic communication was first reported to be successfully used in commercial fiber optic channel to get higher Bit Transmission Rate (BTR) [Apostolos *et al.* (2005)], and included in the latest international standard, i.e., IEEE 802.15.6, for local and metropolitan area networks [802.15.6-2012 (2012)], the chaotic communication research has gained increased attention from the industry community [Bai *et al.* (2018); Kaddoum (2017)]. There are three fundamental discoveries, which show that the chaotic baseband signals have better wireless communication performance as compared with the conventional system. Firstly, the information entropy of the received signal is unaltered after the chaotic signal is transmitted via the wireless channel [Ren *et al.* (2013)], which shows that the chaotic signals can be used in practically wireless channels. Secondly, chaotic waveform is optimal in the sense that the very simple matched filter maximizes the signal to noise ratio (SNR) at the receiver [Corron and Blakely (2015); Ren *et al.* (2016)]. Thirdly, the chaos property of invariant Lyapunov spectrum over wireless channel [Ren *et al.* (2013)] can be used to resist to the inter-symbol interference (ISI) caused by multipath propagation [Yao *et al.* (2017)]. Most importantly, the replacing of the conventional shape-forming filter and the matched filter with the chaotic ones is compatible with the conventional wireless communication hardware; just using even simpler algorithm as compared to conventional one does achieve lower Bit Error Rate (BER).

7.2 Wireless Communication System Configuration

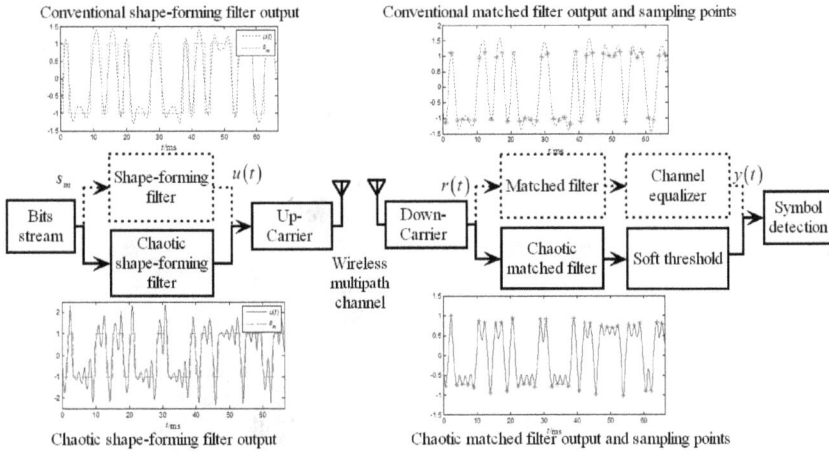

Figure 21 Block diagram of the wireless communication system based on chaotic baseband waveform.

The proposed wireless communication system uses a chaotic shape-forming filter (CSFF) and the corresponding matched filter to replace the conventional ones, as shown in Fig. 21. Besides these, channel equalization in the conventional system is replaced with a soft threshold.

7.3 CSFF and the Corresponding Matched Filter

The CSFF is given by

$$\ddot{u}(t) - 2\beta\dot{u}(t) + \left(\omega^2 + \beta^2\right)(u(t) - s_m) = 0, \tag{7.1}$$

where $u(t)$ is the chaotic baseband signal, $s_m \in \pm 1$ is the binary information symbol sequence to be transmitted, $\omega = 2\pi f$, $\beta = f\ln 2$, and f is the base frequency of the chaotic signal. Signal $u(t)$ is proved to be chaotic with positive Lyapunov exponent $\lambda = \beta$ [Corron *et al.* (2010)]. The shape-forming filter admits a basis function, $p(t)$, given by

$$
p(t) = \begin{cases} \left(1 - e^{-\frac{\beta}{f}}\right) e^{\beta t} \left(\cos(\omega t) - \frac{\beta}{\omega} \sin(\omega t)\right), & t < 0 \\ 1 - e^{\beta\left(t - \frac{1}{f}\right)} \left(\cos(\omega t) - \frac{\beta}{\omega} \sin(\omega t)\right), & 0 \le t < \frac{1}{f} \\ 0, & t \ge \frac{1}{f} \end{cases} \quad (7.2)
$$

The corresponding matched filter is used to maximize the SNR, as shown by

$$
\ddot{y}(t) + 2\beta \dot{y}(t) + \left(\omega^2 + \beta^2\right) (y(t) - r(t)) = 0, \quad (7.3)
$$

where $r(t)$ is the filter input, which is the received chaotic baseband signal after down-carrier, and $y(t)$ is the filter output. The CSFF and the corresponding matched filter can be implemented using the simple electronic circuit [Cohen and Gauthier (2012); Ren *et al.* (2016)] or digital Finite Impulse Response (FIR) Filter [Yao *et al.* (2018)].

7.4   Inter-symbol Interference (ISI) Resistance Technique

A proper threshold, θ, can be designed to resist the ISI [Yao *et al.* (2017)]. The sampling point, y_n, of the matched filter output is given by

$$
y_n = \sum_{l=0}^{L-1} s_n C_{l,0} + \underbrace{\sum_{l=0}^{L-1} \sum_{m=-\infty}^{n-1} s_{n+m} C_{l,m}}_{I_{past}} + \underbrace{\sum_{l=0}^{L-1} \sum_{m=n+1}^{\infty} s_{n+m} C_{l,m}}_{I_{future}} + W, \quad (7.4)
$$

where $W = \int_{-\infty}^{\infty} p\left(\tau - \frac{n}{f}\right) n(\tau) d\tau$ is the noise effect, α_l and τ_l are the attenuation coefficient and delay time for the lth multipath in wireless channel, $n(t)$ is the Gaussian noise, $I_{past} + I_{future}$ is the interference caused by both the past and future symbols, and $C_{l,j}$ is given by

$$
C_{l,j} = \begin{cases} \alpha_l D \left(2 - e^{-\beta/f} - e^{\beta/f}\right) \begin{pmatrix} A\cos(\omega(\tau_l/f)) \\ + B\sin(\omega(\tau_l/f)) \end{pmatrix}, & |\tau_l + j/f| \ge 1/f \\ \alpha_l \begin{cases} A\left(D\left(2 - e^{-\beta/f}\right) - D^{-1}e^{-\beta/f}\right) \cos(\omega(\tau_l/f)) \\ + B\left(D\left(2 - e^{-\beta/f}\right) + D^{-1}e^{-\beta/f}\right) \sin(\omega(\tau_l/f)) \\ + 1 - |\tau_l + j| \end{cases}, & else \end{cases}
$$

$$(7.5)$$

where $A = \frac{\omega^2 - 3(\beta/f)^2}{4\beta\left(\omega^2 + (\beta/f)^2\right)}$, $B = \frac{3\omega^2 - (\beta/f)^2}{4\beta\left(\omega^2 + (\beta/f)^2\right)}$, $D = e^{-\frac{\beta}{f}|\tau_l + j|}$.

Due to the future symbols cannot be predicted at the current time in practical situations, $\theta = I_{past}$ is chosen as the suboptimal threshold. The nth decoded information bit, s_n', is given as

$$s_n' = \begin{cases} +1, \ y_n > \theta \\ -1, \ y_n \leq \theta \end{cases}, \tag{7.6}$$

where initial threshold $\theta = 0$ is used to decode the symbol.

7.5 Simulation Result

The simulation result over multipath wireless channel is given in Fig. 22, where the conventional communication scheme employs Root Raised Cosine shape-forming filter and Binary Phase Shift Keying (BPSK) with Minimum Mean Square Error (MMSE) equalizer. In Fig. 22, the down triangular mark and the dot mark lines represent the BER of the BPSK with and without the channel equalization, respectively. The diamond mark, up triangular mark and the rectangular mark represent the BER of the proposed method using $\theta = 0$, $\theta = 0$ with the MMSE equalization, and $\theta = I_{past}$ without equalization, respectively. It can be seen from Fig. 22, the simulation results show that the BER of Chaos with $\theta = I_{past}$ is lower than that of the BPSK with MMSE, demonstrating the better BER performance in the wireless communication channel.

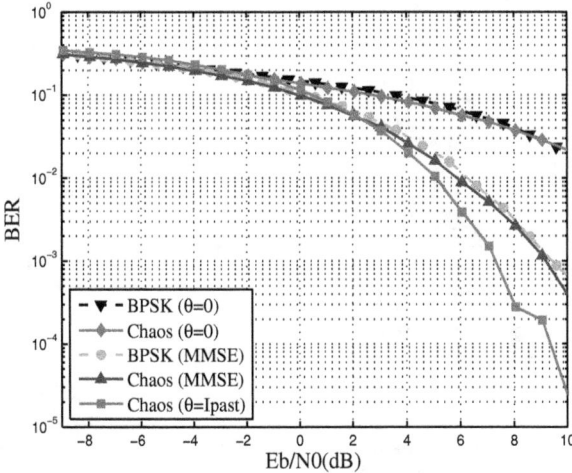

Figure 22   Simulation performance comparison of chaotic/conventional communication in two-path channel, where the base frequency $f = 0.6kHz$, delay time $\tau = [0 \ 1.67\text{ms}]$, and attenuation coefficient $\alpha_l = [1 \ 0.6]$.

7.6   Conclusion

A wireless communication scheme is proposed using a chaotic shape-forming filter and the corresponding matched filter to replace the conventional ones. It is compatible with the traditional wireless communication system hardware. The ISI effect is decreased using chaos property to achieve lower BER under wireless multipath channel with simpler software algorithm. The work reports a practical way to use chaos in wireless communication system to improve the performance.

SDE implementation of chaos-based communications systems

Géza Kolumbán

Pázmány Péter Catholic University, Budapest, Hungary

8.1 Introduction

Since the discovery of chaotic phenomenon, a lot of areas have been proposed for its application. However, these proposals have been disregarded because the feasibility and benefits of chaos-based systems have been verified only by computer simulations and not in real application scenarios.

Chaotic signals are not used in conventional systems, therefore, off-the-self chaotic circuits and blocks are not available. To verify the feasibility and to prove the advantages of a chaos-based solution, each chaotic circuit, block and IC should be implemented from scratch. However, the chaotic community has never had the resources and knowledge to implement the building blocks of chaos-based systems.

Software Defined Electronics (SDE) offers a solution to this problem provided that band-pass signals carry the information. In SDE, all RF/microwave/optical band-pass signals are converted into BaseBand (BB) by a universal HW transformer and every application is implemented entirely in SW running in baseband. Because of the SW implementation, even a computer simulator used in the research phase can be turned directly into a working chaos-based system being capable of generating and processing all real-world physical signals.

The basic idea of SDE concept and the main steps of SDE implementation of a chaos-based FM-DCSK radio link are summarized here. References to all missing details are provided among the references.

8.2   Concept of Software Defined Electronics

RF band-pass signals are processed in ICT and measurement systems. In SDE, every application is implemented entirely in SW and universal RF HW transformers are used to establish the transformation between the analog physical signals measured in real-world and the data sequences processed and generated in baseband on a computing platform. Block diagram of SDE concept is depicted in Fig. 23 where the RF band-pass analog signals and the BB data sequences are shown in red and blue, respectively. Note, the SDE concept is a generalization of software defined radio [Mitola (1992)], where SDE can be used to implement any kind of band-pass systems.

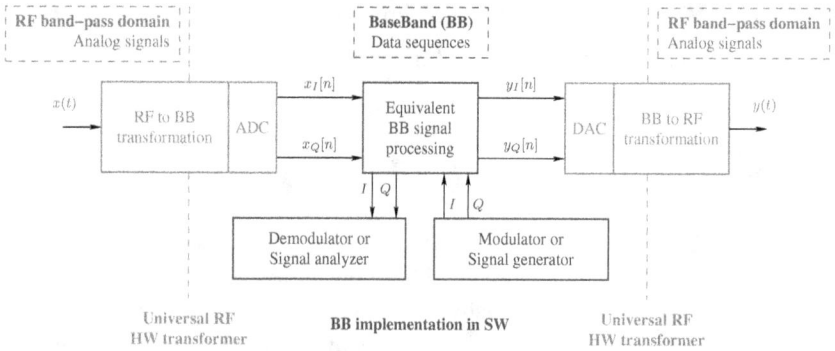

Figure 23   Block diagram of equivalent BB implementation. Conversions between the RF band-pass and BB low-pass domains are performed by the universal HW transformers.

To implement an application in SW, the real-world analog signals have to be digitized. The key issue is the use of minimum sampling rate where the information carried by the analog signal cannot be corrupted or distorted by sampling. This lowest sampling rate attainable theoretically can be achieved by using the theory of complex envelopes and equivalent baseband transformation.

The transformation between the analog and digital domains is performed in both directions by universal HW transformers shown in green in Fig. 23. These devices are universal because the same transformers are used in each application and the implementation of a new application needs to change only the SW used in BB (see blue part of Fig. 23). The universal HW transformers performs the transformation between the (i) RF band-pass and BB low-pass domains and (ii) ADC or DAC conversion. They are available as ICs, USRP devices or PXI/PXIe testbeds [Kolumbán (2018)].

8.3   Mathematical Background

To get the minimum sampling rate, the RF band-pass signal $x(t)$ is decomposed into a product of a complex envelope $\tilde{x}(t)$ and center frequency ω_c

$$x(t) = \Re\left[\tilde{x}(t)\exp(j\omega_c t)\right]$$

where $\Re$ is the real-part operator [Haykin (1994)]. As shown in Fig. 23. only the complex envelopes are processed in BB, their real and imaginary parts are referred to as in-phase (I) and quadrature (Q) components

$$\tilde{x}(t) = x_I(t) + jx_Q(t).$$

Because the center frequency has been removed, the complex envelope is a low-pass signal and its bandwidth is equal to the half of the bandwidth $2B$ of RF band-pass signal. Note, the required sampling rate in BB does not depend on ω_c, it is determined exclusively by B. In general $B << f_c$.

BB equivalents are available for (i) deterministic signals, (ii) LTI blocks and (iii) random processes, that is, for all constituting elements of a linear system. Characteristics of BB equivalents are as follows:

- a BB equivalent can be derived for every band-pass linear system;
- all BB equivalents have a low-pass property;
- BB equivalents allow to use the minimum sampling rate attainable theoretically. It is determined by the half of RF bandwidth;
- RF band-pass analog signal processing can be fully substituted by an equivalent digital one performed in BB;
- except ω_c, the BB equivalent retains all information available in the RF band-pass domain;
- it is a representation and not an approximation, consequently, distortion does not occur.

8.4   Derivation of BB Equivalents

A step-by step process has been developed for the derivation of BB equivalents [Kolumbán et al. (2012)]. To illustrate the goal of the step-by-step derivation its input and output are shown here, that is, the block diagrams of an FM-DCSK telecommunications system are given in both the RF band-pass and BB domains.

As depicted in Fig. 24, the real-world FM-DCSK radio link includes a modulator, radio channel, channel filter and demodulator. In the figure the

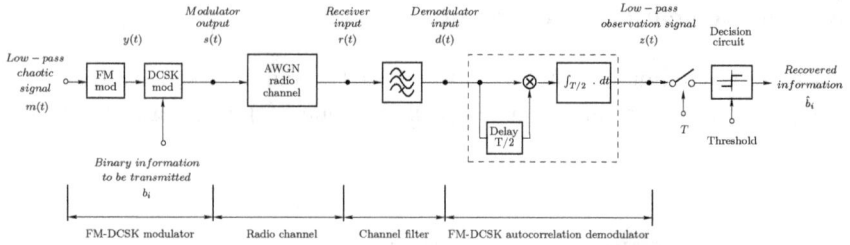

Figure 24   RF block diagram of an FM-DCSK radio link.

low-pass chaotic signal is denoted by $m(t)$, b_i is the binary information to be transmitted and $z(t)$ denotes the low-pass observation signal which is fed into the decision circuit to recover the transmitted information. Except the chaotic and observation signals, all signals plotted in the RF model are RF band-pass signals. During the derivation of BB equivalent, all RF band-pass signals have to be substituted by their complex envelopes.

The BB equivalent of FM-DCSK radio link derived in [Kolumbán (1998)] is depicted in Fig. 25. The upper and lower arms process the in-phase and quadrature components of complex envelopes, respectively. To get a simple BB equivalent which is easy to interpret, only the analog signals are shown in Fig. 25. The BB equivalent run on a computing platform uses the digitized versions of these signals.

The functionality of each block of an FM-DCSK radio link is given in Fig. 24 where, beyond the low-pass chaotic and observation signals, all RF band-pass signals are identified. The in-phase and quadrature components of their complex envelopes are given in Fig. 25, where the BB equivalent of each RF block is also shown. This BB equivalent is implemented in SW.

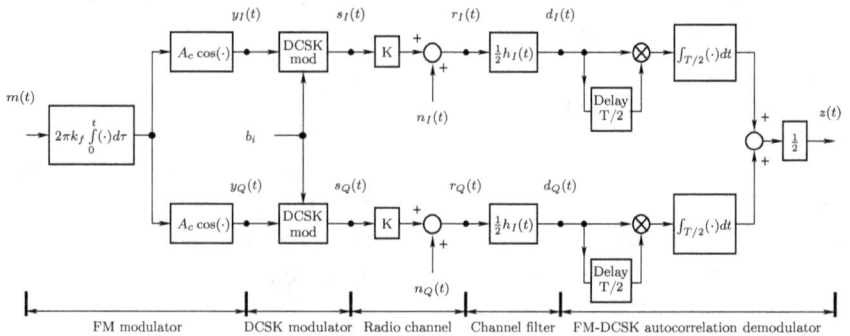

Figure 25   BB equivalent of the FM-DCSK radio link.

8.5   Application of SDE Concept in Scientific Research

Photo of the PXIe-based universal RF SDE platform can be seen in Fig. 26 where a PXIe chassis [on the top-left, (1)], a monitor providing the GUI for the SW implementation in BB [bottom-left, (2)] and a microwave stand-alone spectrum analyzer [right side, (3)] are shown.

The 2.4-GHz FM-DCSK transmitter and receiver are implemented in SW running on the PXIe embedded controller in BB. The SW GUI serves as an interface to the equivalent BB implementation, therefore, each parameter, waveform and spectrum visualized on the monitor are in baseband. The generated FM-DCSK data sequence is converted into the RF domain by the Tx HW transformer and radiated into the air by the Tx antenna. The received signal is picked up by the Rx antenna and converted into BB data by the Rx HW transformer. The stand-alone microwave spectrum analyzer checks the spectrum in the real-world RF radio channel.

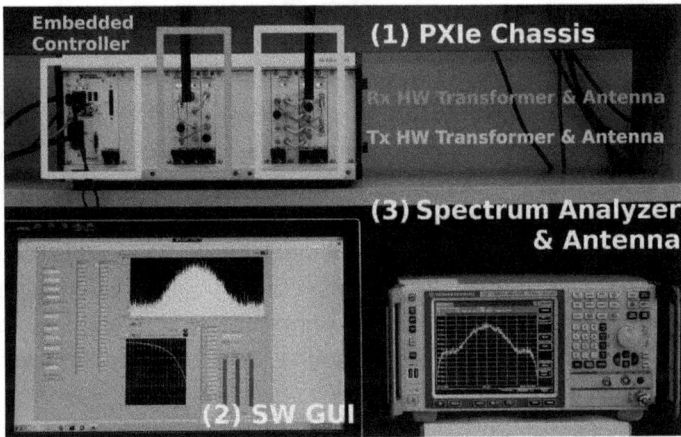

Figure 26   Photo of the PXI-based universal RF SDE platform. The Rx and Tx transformers plugged into the PXIe chassis perform the transformation between the Rx ⇒ BB and BB ⇒ RF domains, respectively.

Acknowledgements

This work was an EFOP project entitled *Thematic Research Cooperation Establishing Innovative Informatic and Info-Communication Solutions*. It was financed by the European Union and co-financed by the European Social Fund under grant number EFOP-3.6.2-16-2017-00013.

Stability analysis of sparse binary neural networks

Seitaro Koyama, Shunsuke Aoki and Toshimichi Saito

Hosei Univerisity, Japan

9.1 Introduction

We introduce the dynamic binary neural network (DBNN [Sato and Saito (2017)]) and consider its basic dynamics. The DBNN is a recurrent type artificial neural network characterized by signum activation function [Gray and Michel (1992)], ternary connection parameters, and integer threshold parameters. Depending on the parameters, the DBNN can generate various binary periodic orbits. The DBNN has advantages in precise numerical analysis [Koyama *et al.* (2018)] and hardware implementation [Aoki *et al.* (2018)] as compared with artificial neural networks with real connection parameters and smooth activation function. In order to store a desired target periodic orbit (TBPO), we have a learning method based on a correlation learning and a condition of parameters for storage of a class of TBPOs. The TBPO is applicable to various engineering systems including control signal of switching circuits [Bose (2007)], sound data encoders [Wada *et al.* (2002)], and associative memories [Jiang *et al.* (2016)].

When a TBPO is stored into a DBNN, stability of the TBPO is an important problem. In order to characterize the stability, we introduce a simple feature quantity. Using the feature quantity, we consider effects of connection sparsity on the stability. Performing numerical experiments for a typical TBPO, we can suggest that, as the connection sparsity increases, the stability of the TBPO can be reinforced. There exists suitable sparsity in which the stability is very strong. As the connection sparsity increases, power consumption in hardware can be reduced as compared with full binary connection. As the stability becomes stronger, reliability and robustness can be reinforced. Our results provide basic information for realization of efficient hardware for useful engineering applications.

9.2   Dynamic Binary Neural Networks

The DBNN dynamics is described by the following difference equation:

$$x_i^{t+1} = F\left(\sum_{j=1}^{N} w_{ij} x_j^t - T_i\right), \quad F(x) = \begin{cases} +1 \text{ if } x \geq 0 \\ -1 \text{ if } x < 0 \end{cases} \tag{9.1}$$

where $i \in \{1, \cdots, N\}$ and $j \in \{1, \cdots, N\}$ and N is the number of cells. The $x_i^t \in \{-1, +1\}$ is the i-th binary state at discrete time t. The connection parameters are ternary $w_{ij} \in \{-1, 0, +1\}$ and the threshold parameters are integer $T_i \in \{-N, -N+1, \cdots, 0, 1, \cdots, N+1\}$. Let $\boldsymbol{x}^t \equiv (x_1^t, \cdots, x_N^t)$. We abbreviate Eq. (9.1) by $\boldsymbol{x}^{t+1} = F_D(\boldsymbol{x}^t)$. As an initial state $\boldsymbol{x}^1$ is applied, the DBNN generates a sequence of binary vectors $\{\boldsymbol{x}^t\}$.

We try to visualize the dynamics. The domain of the DBNN is a set of binary vectors $\boldsymbol{B}^N$ that is equivalent to a set of points $L_D = (C_1, \cdots, C_{2^N})$, $C_i \equiv i/2^N$. Hence the DBNN $\boldsymbol{x}^{t+1} = F_D(\boldsymbol{x}^t)$ is equivalent to the digital return map (Dmap) $\theta^{t+1} = f_D(\theta^t)$ where $\theta \in L_D$ corresponds to the decimal expression of $\boldsymbol{x} \in \boldsymbol{B}^N$. Figure 27 shows examples of DBNN and Dmap.

Since the number of the points in L_D is finite, the steady state of the Dmap must be a periodic orbit. Here we give basic definitions of the periodic orbit. A point $\theta_p \in L_D$ is said to be a periodic point with period p if $f_D^p(\theta_p) = \theta_p$ and $f_D(\theta_p)$ to $f_D^p(\theta_p)$ are all different where f_D^k is the k-fold composition of f_D. A sequence of the periodic points, $\{f_D(\theta_p), \cdots, f_D^p(\theta_p)\}$, is said to be a periodic orbit (PEO). A PEO with period 6 is shown in Dmap in Fig. 27(b). A point $\theta_q \in L_D$ is said to be an eventually periodic point (EPP) with step q if θ_q is not a periodic point but falls into some PEO after q steps: $f_D^q(\theta_q) = \theta_p$ where θ_p is some PEP. Since a PEO is equivalent to a BPO of the DBNN, let the term BPO mean both BPO and PEO hereafter.

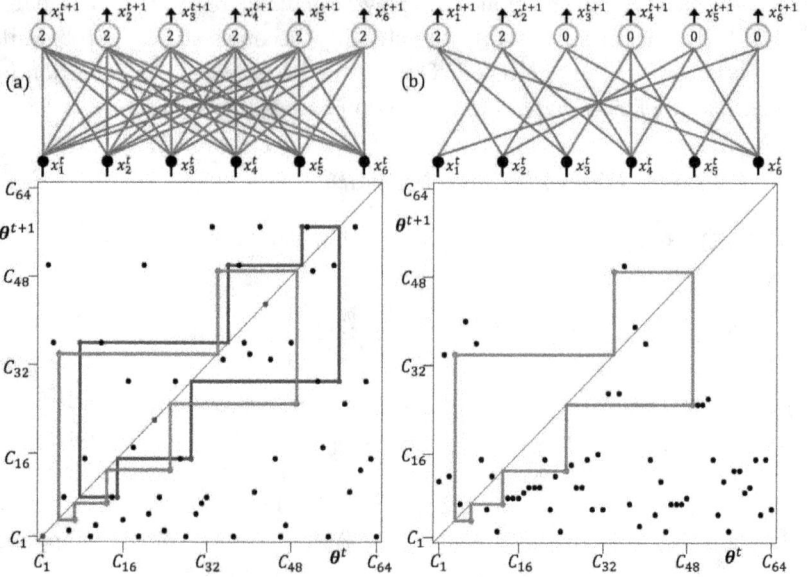

Figure 27   DBNN (red branch: $w_{ij} = +1$, blue branch: $w_{ij} = -1$, green circle: T_i) and Dmap (red: TBPO, blue: spurious BPOs, black: EPPs). (a) $W = W_1$ in Eq. (9.5). SR=0, $\alpha = 6/58$. (b) $W = W_2$ in Eq. (9.5). SR=18/30. $\alpha = 58/58$.

9.3   Orbit Stability versus Connection Sparsity

We consider stability of a target binary periodic orbit (TBPO). The TBPO with period p is defined by

$$z^1, \cdots, z^p, \cdots, \quad \begin{cases} z^t = z^s \text{ for } |t - s| = np \\ z^t \neq z^s \text{ for } |t - s| \neq np \end{cases}, \quad z^t \equiv (z_1^t, \cdots, z_N^t)^T \in \boldsymbol{B}^N$$

(9.2)

where n denotes positive integers. A TBPO can be stored into the DBNN if the following storage condition is satisfied [Koyama *et al.* (2018)].

$$L(i) < T_i \leq R(i) \ \forall i, \quad \begin{cases} R(i) = \min_\tau (\sum_{j=1}^N w_{ij} z_j^\tau) \text{ for } F_D(z_i^\tau) = +1 \\ L(i) = \max_\tau (\sum_{j=1}^N w_{ij} z_j^\tau) \text{ for } F_D(z_i^\tau) = -1 \end{cases}$$

(9.3)

As a typical example, we use a TBPO with period 6 for $N = 6$:

$$z^1 = (-1, -1, -1, -1, +1, +1)^\intercal \quad z^2 = (+1, -1, -1, -1, -1, +1)^\intercal$$

$$z^3 = (+1, +1, -1, -1, -1, -1)^\intercal \quad z^4 = (-1, +1, +1, -1, -1, -1)^\intercal \quad (9.4)$$

$$z^5 = (-1, -1, +1, +1, -1, -1)^\intercal \quad z^6 = (-1, -1, -1, +1, +1, -1)^\intercal$$

This TBPO is applicable to a switching signal of a basic AC/DC inverter that converts a 3-phase AC input into a DC-like output via 6 switches [Sato and Saito (2017)]. Using the following parameters, the TBPO can be stored into the DBNN.

$$
W_1 = \begin{pmatrix}
+1 & -1 & -1 & -1 & +1 & +1 \\
+1 & +1 & -1 & -1 & -1 & +1 \\
+1 & +1 & +1 & -1 & -1 & -1 \\
-1 & +1 & +1 & +1 & -1 & -1 \\
-1 & -1 & +1 & +1 & +1 & -1 \\
-1 & -1 & -1 & +1 & +1 & +1
\end{pmatrix}
\quad
W_2 = \begin{pmatrix}
0 & -1 & -1 & 0 & 0 & +1 \\
+1 & 0 & -1 & -1 & 0 & 0 \\
0 & +1 & 0 & -1 & 0 & -1 \\
0 & 0 & +1 & +1 & -1 & 0 \\
0 & -1 & 0 & +1 & 0 & -1 \\
-1 & 0 & 0 & 0 & +1 & +1
\end{pmatrix}
\quad (9.5)
$$

$$
T_1 = \begin{pmatrix} 2 & 2 & 2 & 2 & 2 & 2 \end{pmatrix}^\mathsf{T}
\quad
T_2 = \begin{pmatrix} 2 & 2 & 0 & 0 & 0 & 0 \end{pmatrix}^\mathsf{T}
$$

The matrix W_1 is given by binarization of the connection matrix from the correlation learning [Koyama *et al.* (2018)].

$$
w_{ij} = \begin{cases} +1 \text{ for } c_{ij} \geq 0 \\ -1 \text{ for } c_{ij} < 0 \end{cases}, \quad c_{ij} = \sum_{\tau=1}^{p} z_i^\tau z_j^{\tau+1}
\quad (9.6)
$$

The connection matrix W_2 is given by a heuristic sparsification of W_1. In order to consider stability of the TBPO, we introduce a feature quantity

$$
\alpha = \frac{\text{The number of EPPs falling into a TBPO}}{2^N - p}, \quad 0 \leq \alpha \leq 1.
\quad (9.7)
$$

As α increases, stability of the TBPO increases. If $\alpha = 1$ then all the initial points fall into the TBPO. The connection sparsity is characterized by

$$
\text{SR} = \frac{\text{The number of zeros in } W}{N^2 - N}, \quad 0 \leq \text{SR} \leq 1.
\quad (9.8)
$$

As SR increases, the sparsity increases where each input neuron is assumed to connect to at least one output neuron. For simplicity, we consider the case $N = 6$ and fix one element for each column/row:

$$
w_{21} = w_{32} = w_{43} = w_{54} = w_{65} = w_{16} = +1.
$$

If all the other elements of W are zero, then SR= 30/30 and the DBNN is equivalent to the shift register where all the points are periodic points ($\alpha = 0$). DBNNs and Dmaps in Fig. 27 correspond to Eq. (9.5): SR=0, $\alpha = 6/58$ for W_1 and SR=18/30. $\alpha = 58/58$ for W_2.

Inserting zero element successively to the full binary connection W_1, α of the TBPO is calculated. First, we insert one zero element into all the possible position of the W_1 (SR= 1/30) and calculate α. For each value of SR (e.g. 17/30), we add one zero element into W, calculate α, and repeat until SR= 30/30. Figure 28(a) shows histogram of the DBNNs into

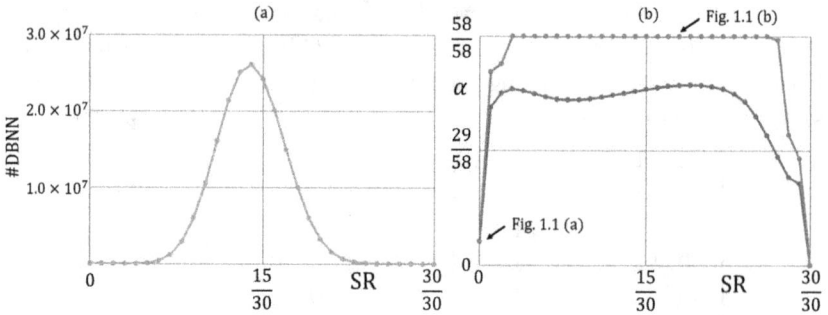

Figure 28   Storage and stability. (a) The number of DBNNs into which the TBPO is stored. (b) Global stability. Red and blue plots denote maximum and average of α.

which the the TBPO is stored. We have confirmed that the TBPO can be stored into 191102976 DBNNs out of 2^{30} DBNNs. Figure 28(b) shows the stability (α) of the TBPO for SR in the 191102976 DBNNs. We can see that as SR increases, α increases and reaches the optimal value $\alpha = 1$ where the stability of TBPO is the strongest. As SR increases further, α varies and decreases to the minimum value $\alpha = 0$ for SR= 30/30.

9.4   Conclusions

The DBNN is introduced and stability of a TBPO is considered. Using a typical TBPO example, effects of the connection sparsity on stability of the TBPO is investigated. The result suggests existence of suitable connection sparsity where stability of the TBPO is the strongest. In order to develop the study of DBNN, we should consider various problems including stability analysis of various TBPOs, stabilization of desired TBPOs, and FPGA based hardware implementation for engineering applications. An elementary FPGA based hardware design can be found in [Aoki *et al.* (2018)].

Memory architectures with cryptographic features on board to support a secure control system

Antonino Mondello

Micron NVE Principal Design Engineer

10.1 Introduction

For many years the system engineers focus was the safety, efficiency, performance and robustness of control systems. They adapted the design of the controller to the technological developments introducing in the field the latest progress of all the engineering branches. Since the introduction of microprocessors, they were used as calculus engine for the control systems and in the field, appears the new controller architecture depicted in Fig. 29.

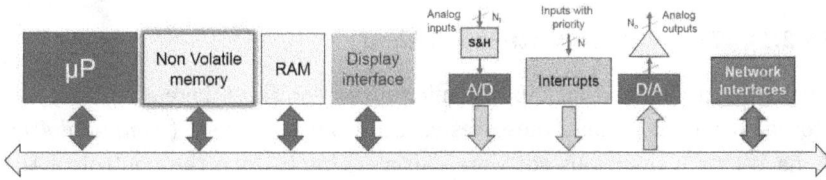

Figure 29   New controller architecture.

This new architecture was a leap forward for the control field, with it is born the *digital control systems* era. The major benefit of this architecture is the possibility to easily change control parameters and control algorithms. On the contrary this innovation introduced a severe safety issue because an un-authorized person could be able to change intentionally or accidentally

51

the control parameters and then the behavior of the system controlled. For many years the only approach followed to avoid this kind of issue was the control system segregation. This protection became weak year after year due to the extension of the system connectivity (see red block in Fig. 29). Today, in the internet of things (IoT) era, the control systems must be interconnected each other, by public network infrastructures, to implement Distributed Control System (DCS) and also to allow on-field system re-configuration and updates. Consequently, control system engineers must focus also to the security aspect to mitigate the hacking risks. The aim of this work is the introduction, in the system design, of some cryptographic concepts to reduce the controller vulnerability.

10.2   Safety and Security Treats in a Controller

Even if the microprocessor is the heart of the system, the most sensitive components, in terms of security, are the storage devices especially the non-volatile ones. This because all the information that describes the control system and the control algorithms are stored there and any attempt to change the controller functionality consists in their alteration. Hence, the storage device cannot be *a simple* container of data, but, also, it must be resilient to some different types of treats: *Unauthorized use of the storage device, storage device replacement, storage device content changes.* A storage device, resilient to such treats, must implement the security features described in the following paragraphs.

10.2.1   *The Authentication Problem*

In the controller of Fig. 29, the microprocessor (host hereafter) and the storage device exchanges messages chosen from a finite set (*command set*). A hacker or a malicious software can attempt to take the control of the board for evil purposes. This problem can be fixed by implementing an authenticated access to the storage device, by using a cryptographic prim-itive called Message Authentication Code (MAC) [Stinson (2005)] used to calculate, what is called, **signature**. Only the *authentic* host and the stor-age device are able to calculate the signature because the calculus is based on a *secret key* that they previously shared in a secure environment. The storage device will execute the command request, if the associated signa-ture matches the one calculated by itself using its copy of secret key. The security strength of this method, is not based on the secrecy of the MAC function, but, on the impossibility to calculate the right message signature without knowing the secret key, and, to guess the secret key from a given signature (*Kerckhoffs principle*). The command signature is used in the following way to ensure a robust system authentication:

(1) **The host:** Sends the commands to be executed and the related signature calculated, by using the secret key, with the formula: $Signature = MAC(Secretkey, Command)$

(2) **The storage device:** Takes, from the received message, the command field and locally calculates the signature by using the same above formula with its copy of the secret key, then compares the signature calculated with the one received; if they match the command is executed otherwise it is discarded; The storage device sends the command result and the related signature to the host

(3) **The host:** Verifies the authenticity of the received message by locally calculating the signature, with its secret key copy, and comparing it with the one received.

This approach is very robust in guaranteeing authentication, but it has a problem the *replay attack*: *if someone intercepts the message, by using a bus sniffer, they can reuse it later* because it is correctly signed by the authorized host. A solution to this problem can be found by adding in the message packet a field, called *freshness*, that changes at each command transaction. So, the above protocol is changed in this way: the receiver checks if the freshness in the message is the expected one, and then, verifies if the signature received matches the local one given by the formula: $Signature = MAC[Secretkey, (Command|Freshness)]$. The freshness ensures that on the transaction of two identical commands, the related signature is always different, so there is no way to reuse twice the same command. There are many ways to define the freshness: it can be the transmission time-stamp, or more simply, a number that is incremented at each transition and with the property to be monotonic. The freshness approach is allowed by the robustness of the MAC algorithm, in fact, despite only a bit is changed in the message, the related signature is unpredictable.

10.2.2   *Component Identity Proof*

An extreme technique *to hack* the controller consists in the storage component replacement with a non-genuine one. To detect this event, the storage device can be able to prove its identity on-demand by host, providing: a component unique identification code (UID), the current value of freshness and the related signature: $Signature = MAC[Secretkey, (UID|Freshness)]$. Because only the genuine component knows the secret key, the set: {UID, Freshness, signature} represents a certification of genuineness.

10.2.3   *The Data Attestation Problem*

The storage devices must guarantee the genuineness of data stored. After each controller initialization or on-demand by host, the integrity of the data stored must be checked. In case of data corruption, due to a malicious or a physical problem, it must restore the data from a genuine source to guarantee the controller functionality in all the conditions. The integrity check can be done by using another cryptographic primitive called HASH. HASH is able to process the stored data providing *a short* pattern called digest. Once a digest is calculated on a genuine data pattern, the result (*golden digest*), is stored in an area not user-accessible. When requested, the current digest is compared with the golden one, to detect any intentional or accidental content modifications.

10.3   MAC and HASH Functions

There are several functions that can be used as MAC functions to generate the signature of a message; these functions must satisfy some important requirements:

(1) *Easy* to be calculated: Given a message X, the calculus of $MAC(key, X)$ is not computationally difficult;

(2) *Hard* to be inverted: Given a $MAC(key, X)$, it must computationally infeasible to determine the key value by knowing X and the MAC value;

(3) *Negligible* collision probability: Given two messages $X \neq Y \Rightarrow MAC(key, X) \neq MAC(key, Y)$.

A powerful MAC function used in many cryptographic systems is the HMAC-SHA256 described in [Stinson (2005); FIPS (2015a)]. A function is eligible as a HASH function if it satisfies certain requirement similar to the MAC ones, it must be: *Easy* to be calculated, *Hard* to be inverted and must have *Negligible* collision probability. A powerful HASH function is the SHA256 described in [Stinson (2005); FIPS (2015b)]. Both SHA256 and

HMAC-SHA256 can process messages up to 2^{64} bits in length and produce a 256-bit signature. As of today, there is no known possibility to invert these functions and no collision conditions for them, have been identified.

10.4 The Secure Storage Device

The usage of a MAC function and a freshness mechanism applied to the storage device commands allow to define what is called a **secure command set**. In fact, if all the commands used for data modification like: program, erase, require freshness and signature fields, then the storage device is usable only by whom is knowing the secret key. Moreover, adding a command (*Request_UID*) able to ensure component identity, as described in Sec. 10.2.2, and another one (*Data_measure*) able to provide data attestation, as described in Sec. 10.2.3, permit to define the component as **secure storage device**.

10.5 A Controller Implementation

A secure storage device can be used as brick to build a secure system controller. The control engineer designers can store all the sensitive data and codes inside a genuine storage device (Sec. 10.2.2) and verify them before their usage by the *data_measure* command (Sec. 10.2.3). In case of data corruption, a genuine copy is recovered from a user inaccessible area, or, downloaded from the cloud by using the network controller interface. To implement a classical Observer-Predictor (OP) [Levine (2000)] control scheme, see Fig. 30, all the matrices values used to implement the predictor, the observer and the control block are stored, in a specific area. The code that implement the algorithms able to solve/iterate the control equations are stored in another dedicated area. Similarly, to implement an artificial neural network (ANN) based controller, see Fig. 31, all the neuron biases

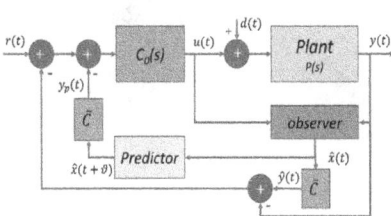

Figure 30   Observer-Predictor control.

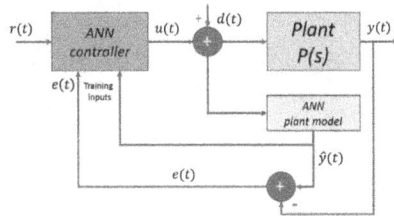

Figure 31   Neural network control.

and synaptic weights are stored in a specific area, and the code, that implement the ANN training algorithms, see [Hagan *et al.* (1996)], and updates all the neurons, are stored in another one. In conclusion, the usage of the cryptography provides huge benefits in terms of safety and security of the controller and guaranteeing, at the same time, high flexibility in controller update by using the authenticated command set.

Acknowledgements

The author wishes to thank the NVE Catania design team for the considerable work performed in designing of secure flash memory device. Special thanks to *Tommaso Zerilli*, Catania NVE senior design manager, for the valuable discussion about this topic.

Evidence of multistability in a multiscroll generator system

J.L. Echenausía-Monroy, J. Hugo García-López, Rider Jaimes-Reátegui
and G. Huerta-Cuellar

*Centro Universitario de los Lagos, Universidad de Guadalajara, Enrique
Diaz de León 1144, Paseos de la Montaña, Lagos de Moreno 47460,
Jalisco, México*

11.1 Introduction

The phenomenon known as multistability is defined, from the point of view
of dynamical systems, as the coexistence of more than two possible final
states, or attractors, for a set of control parameters. The final state to
which the system converges depends, on the initial conditions. This kind of
behavior is common in almost any area of science, like in chemistry [Crowley
and Epstein (1989)], biology [Nickerson (1973)], physics [Kawashima *et al.*
(1991)], economics [Holling (2001)], in the nature itself, among many others
[Pisarchik and Feudel (2014)].

In recent years, the design and control of systems with multiple scrolls
have been subject of interest for the scientific community, locating its origin
in the study of the system modification of Chua's attractor [Chua (1992)],
which possess a double-scroll attractor. This is why it is considered a
multiscroll to all attractor that has at least three scrolls in its phase space.
There are different approaches to obtain this kind of dynamics, as from the
implementation of hysteresis [Arena *et al.* (1995)], the use of sine/cosine
functions, as well as the generation of Piece-Wise Linear functions (PWL)
[Tang *et al.* (2001); Suykens and Huang (1997); Yalcin *et al.* (2000)], to
mention the most studied.

The appearance of multistability in multiscroll systems may be achieved in several ways, i.e., by means of the construction of a particular nonlinear function that modifies the location of the equilibrium points [Gilardi-Velázquez *et al.* (2017)], or by applying some control techniques that generates the coexistence of several monostable attractors [Ontañón-García and Campos-Cantón (2017)]. Here, numerical simulation results of a multiscroll generator system implementing the *round* function are presented, where two main targets are achieved; i) Control in the generation of monostable multiscroll attractors by modifying a single control parameter in the non-linear function, and ii) Obtention of multiscroll coexisting attractors, without the need of making modifications into the system or its construction, by means of the bifurcation parameter.

11.2   Theoretical Background

The studied multiscroll generator system is described by a set of three coupled differential equations, Eq. (11.1), which makes use of the generalization of a PWL function through the implementation of the *round* function as a commutation law.

$$\dot{x} = y,$$
$$\dot{y} = z, \qquad\qquad (11.1)$$
$$\dot{z} = -\alpha_1 x - \alpha_2 y - \alpha_3 z + \alpha_4,$$

where $\alpha_4 = C1 \left[g \left(\frac{x}{C2} \right) \right]$, with $C2 = 0.6$, that defines the size of the scroll. The function $g(x)$ responds to the Nearest Integer Function [Š. (2010)], [Huerta-Cuellar *et al.* (2014)], as follows:

$$g = \left\{ \begin{array}{c} Up\ round\ by\ taking\ \lfloor x + 0.5 \rfloor, \\ Down\ round\ by\ taking\ \lceil x - 0.5 \rceil, \\ Round\ to\ even\ numbers\ to\ avoid\ statistical\ inference. \end{array} \right\}$$

The election of the system control parameters, α_1, α_2, α_3, is based on the combinations of fixed points in a three-dimensional system corresponding to the definition of an Unstable Dissipative System, UDS I [Campos-Cantón *et al.* (2010)]; this classification of dynamical systems responds to a fixed points configuration type *hyperbolic-saddle-node*, with the additional condition that the sum of their eigenvalues must be negative, $\sum \lambda_i < 0$, been λ_i the *ith* eigenvalue of the system, which generates a combination where one eigenvalue is negative real (dissipative component) and the other two are complex conjugated with positive real part (unstable and oscillatory component). This combination in the system eigenvalues are responsible for stretching and successive folding in the dynamic of the system, which favors the generation of multiscroll attractors.

Because of the great variety of combinations of the α parameters, it is better to limit $\alpha_1 = \alpha_2 = \alpha_3$, with the operation zone defined as $0 < \alpha_{1,2,3} < 1$.

11.3   Methodology and Results

The dynamical system, Eq. (11.1), is analyzed by a gradual change of the bifurcation parameter $C1$ for the different $\alpha_{1,2,3}$ values. For this purpose, it is necessary the construction of bifurcation diagrams of local maximums of the state variable $x(t)$, that are calculated by means of randomly changing the initial conditions of the three state variables.

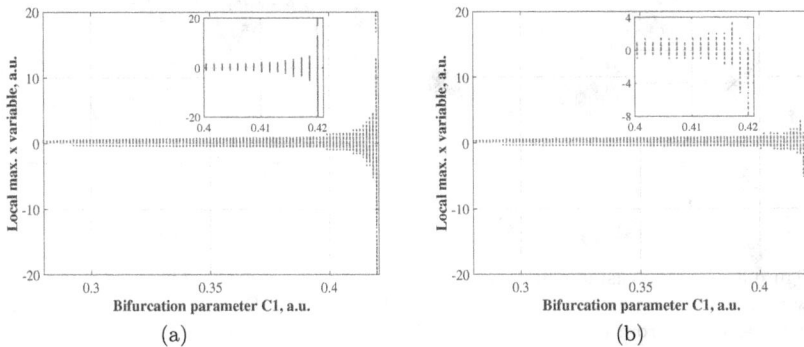

Figure 32   Bifurcation Diagrams (BD) for $\alpha_{1,2,3} = 0.7$ (a) for twenty random initial conditions, (b) for only one initial condition. The insets are a zoom of the corresponding BD showing the multistable region.

By the variation in the bifurcation parameter, $C1$, the system is able to generate from a single-wing attractor, three-five-scrolls, among others. Figure 32 shows the behavior of the system through the modification of the bifurcation parameter for a value $\alpha_{1,2,3} = 0.7$. In Fig. 32(a) it is shown the Bifurcation Diagram (BD) calculated from changing randomly twenty times the initial conditions, meanwhile in Fig. 32(b) it is shown the BD obtained from exploring only one random initial condition. The control in the generation of different monostables states, offers a great flexibility in the system performance. This model allows the generation of a large variety of attractors with different number of scrolls.

As expected, the system presents a dependency between the control parameter and the bifurcation parameter [Echenausía-Monroy *et al.* (2018)]. The generation of monostable attractors occurs for all $C1 < \alpha_1 * C2$. For $C1 = \alpha_1 * C2$, the system presents the transition from a monostable behavior to a multistable dynamic. Figure 32 insets, shows the sensitivity to initial conditions, where a very small change in them can lead to a completely different attractor. Such multistable behavior is shown in Fig. 33, where the system is analyzed for a fixed set of parameters, and only the initial conditions are modified, creating totally different attractors on each change.

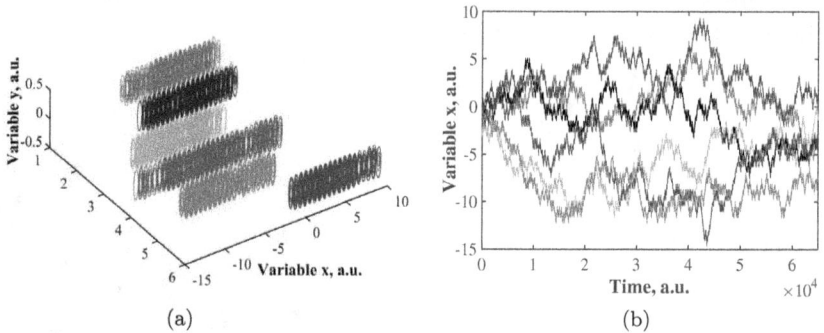

(a)   (b)

Figure 33   (a) Multistable attractors calculated for $\alpha_{1,2,3} = 0.7$, $C2 = 0.6$, $C1 = 0.42$, by randomly changing the initial conditions of the system. (b) Temporal behavior of the state variable x from its corresponding attractor (color) shown in Fig. 33(a).

As can be appreciated in Fig. 33, the differences of the attractors validated the multistable response of the system. Although only the results for one control parameter are shown in here, the behavior described, mono and multistable, it is achieved for every $\alpha_{1,2,3}$ value defined into de UDS I definition.

11.4   Concluding Remarks

In this work, a multiscroll system with the implementation of *round* function has been studied. The parameter $C1$ allows to drive the dynamical response through several monostable attractors until reach a multistable zone, by generating multiscroll coexisting attractors. Since the region of coexistence of attractors is very sensitive to external perturbances or small changes in the initial conditions, it is an interesting task to address, the development of the appropriate control strategies to induce a definite switching between the different present states during the multistable behavior.

Acknowledgements

J.L.E.M. acknowledges CONACyT for financial support (National Fellowship CVU-706850, No. 582124) and the University of Guadalajara, CULagos (Mexico). G.H.C. acknowledges to the University of Guadalajara for sabbatical financial support under the project No. 241596-1.1.9.24. All the authors acknowledge to the University of Guadalajara for financial support under the project *Research laboratory equipment for academic groups in Optoelectronics from CULagos*, R-0138/2016, Agreement RG/019/2016 UdeG, Mexico.

RQA correlations on business cycles: A comparison between real and simulated data

Giuseppe Orlando[1,3] and Giovanna Zimatore[2]

[1] *Università degli Studi di Bari Aldo Moro, Italy*
[2] *Università degli Studi di Camerino, Italy*
[3] *eCampus University, Rome, Italy*

12.1 Introduction

This work concerns the application of recurrence plots and of their quantitative description provided by recurrence quantification analysis (RQA) to appreciate subtle but physiologically relevant changes in the dynamic regime of business time series. RQA aims at a direct and quantitative appreciation of the amount of deterministic structure of time series and has been shown to be an efficient and relatively simple tool in non-linear analysis of a wide class of signals. The technique allows for the identification of sudden phase-changes possibly pointing to mechanistically relevant phenomena. Therefore RQA may be suitable to study business cycles and could be potentially used for early detection of recessions.

The paper is organized as follows. The first Section contains the definition of business cycle and summarizes the literature on recurrence quantification analysis and its applications to economics and finance. The second Section illustrates the methodology and the data. The third Section shows the analysis performed and the results obtained. A final Section draws some concluding remarks, and suggestions for future research.

12.2 Literature Review

According to The National Bureau of Economic Research (NBER) [National Bureau of Economic Research (2008)] a recession is "a significant decline in economic activity spread across the economy, lasting more than a few months, normally visible in real GDP, real income, employment, industrial production, and wholesale-retail sales. A recession begins just after the economy reaches a peak of activity and ends as the economy reaches its trough. Between trough and peak, the economy is in an expansion. Expansion is the normal state of the economy; most recessions are brief and they have been rare in recent decades" (see Fig. 34).

Financial and business cycles in the United States

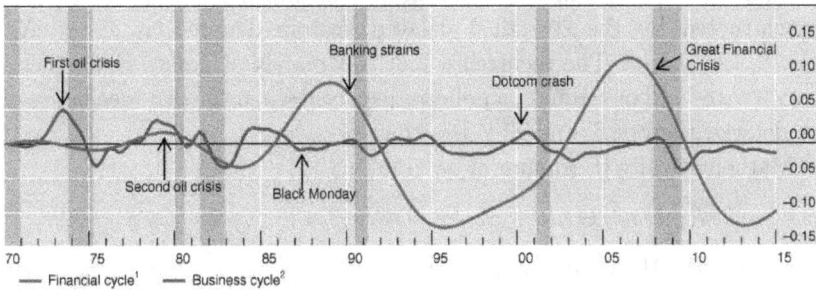

Figure 34   BIS 85th Annual Report 2015.

The ability of RQA to predict catastrophic changes is in line with the fact that RQA is based upon the change in correlation structure of the observed phenomenon known to precede the actual event in many different systems ranging from physiology [Zimatore *et al.* (2011)] and geophysics [Zimatore *et al.* (2017)] to economy [Crowley (2008)], [Chen (2011)], [Moloney and Raghavendra (2012)]. Gorban *et al.* (2010) [Gorban *et al.* (2010)] found out that even before crisis correlation increases as does variance (and volatility) in an economic contest. In particular their dataset composed of the thirty largest companies from the UK stock market within the period 2006–2008 supports the hypothesis of increasing correlations during a crisis and, therefore, that correlation (or equivalently determinism) increases when the market goes down (respectively decreases when it recovers). In [Fabretti and Ausloos (2005)] there were cases where RQA could detect a warning before a crash. In accordance to that [Addo *et al.* (2013)] affirms "the usefulness of recurrence plots in identifying, dating and explaining financial bubbles and crisis". Finally yet importantly [Bastos and Caiado (2011)] found a correlation between RQA measures and the sub-prime mortgage crisis.

12.3   Material and Methods

12.3.1   *Recurrence Quantification Analysis (RQA)*

As mentioned in [Orlando and Zimatore (2018)] the unknown state at the time t of a dynamic system, in the m-dimensional space, can be reconstructed by the so called delayed vectors denoted as $\vec{x}_t = X_t$, $X_{t+d}, ..., X_{t+(m1)d}$. The recurrence plot is a matrix of points (i, j) where, given a threshold or radius ε, a point is displayed when the distance between the delayed vectors $\vec{x}^i$ and $\vec{x}^j$ is less than ε.

Mathematically [Eckmann *et al.* (1987)]

$$R_{i,j}(\varepsilon) = \Theta\left(\varepsilon - \parallel \vec{x}_i - \vec{x}_j \parallel\right) \quad i, j = 1, \ldots, N \quad i \neq j$$

where Θ is the Heaviside function.

RQA defines the overall complexity of the signal in terms of quantitative indices deriving from RP and a range of measures can be obtained such as laminarity, determinism, trend, etc. (for more details see [Orlando and Zimatore (2018)]). RQE computes recurrence quantifications on an epoch-by-epoch basis. Here RQA was carried out on the time series included in the dataset above mentioned (see Sec. 12.3.3) with the following input parameters: Embedding=10; Radius=80; Line=5; Shift=1; Epoch= 50; Euclidean distance=Meandist.

12.3.2   *RQE Correlation Index*

As explained in [Orlando and Zimatore (2017)], let us define the set of indices $\mathcal{I} = [1, n] \subseteq \mathbb{N}$ and, for each $(k, i) \in \mathbb{N}^* \times \mathbb{N}^*$ with $k < n$ and $i \leq n - k + 1$, we define the discrete times for the rolling window $\mathcal{I}_{k,i} = [i, i + k - 1] (\subset \mathbb{N})$ where k and i are, respectively, the size and the window's index. It follows that the number of windows is $q = n - k + 1 (\geq 2)$.

For each $(k, i, l) \in \mathbb{N}^* \times \mathbb{N}^* \times \mathbb{N}^*$ we denote the set of RQE times series as $S = (S^{(l)})_{1 \leq l \leq L}$ where $S^{(l)} = \{S_t^{(l)} \mid 1 \leq t \leq n\}$ represents the l-th RQA measure. Therefore $S^{(l)}$ can be split in q samples according to the rolling window $\mathcal{I}_{k,i}$ such that $S_{k,i}^l = \{S_t^l \mid t \in \mathcal{I}_{k,i}\}$.

For each $l \neq m$, we denote $\rho_{l,m}$ as the Spearman's correlation coefficient between S^l and S^m and $\rho_{l,m}^{k,i}$ as the Spearman's correlation coefficient between $S_{k,i}^l$ and $S_{k,i}^m$. Therefore there are $p = \binom{L}{2}$ pair of correlations $\rho_{l,m}$ and $q \times p$ pair of 'windows' correlations $\rho_{l,m}^{k,i}$ so that the product

$$P_{abs}(\ RQE\ )_{k,i} = \prod_{\substack{l,m=1 \\ l \neq m}}^{L} (1 + |\rho_{l,m}^{k,i}|) \tag{12.1}$$

can be defined as the *RQE (absolute) correlation index* for the rolling window $\mathcal{I}_{k,i}$ and it varies between 0 and 2^p.

12.3.3 The data

Our data consists of the USA GDP time series as detailed in Table 12.1 and of the income (abbreviated to Y) time series as generated by a Kaldor-Kalecki (see [Orlando (2018)], [Orlando (2016)]) (US. Bureau of Economic Analysis, Gross Domestic Product [GDP], Seasonally Adjusted Annual Rate, Percent Change, retrieved from FRED, Federal Reserve Bank of St. Louis `https://fred.stlouisfed.org/series/A191RP1Q027SBEA`, January 3, 2016).

Table 12.1   Time series on income (Y).

#	Time series	Data points	Frequency	Data range (from to)	Type	Account code/ID
1	USA GDP	274	Quarterly	1947-01-01 to 2015-07-01	Y	A191RC1

12.4   Analysis and Results

12.4.1   RQE Correlation Index on GDP Data

In [Orlando and Zimatore (2017)] we investigated whether the above mentioned correlation index can help in understanding the changes in a times series and we found some positive evidence by taking a known random signal normally distributed $\varepsilon \sim \mathcal{N}(\mu, \sigma^2)$ by applying a perturbation to its mean and variance. On that signal we found out that RQE correlation calculated as in Eq. (12.1) is closer than the other and it is able to detect more finely changes in the times series. Here we want to understand if the RQE

correlation index can help in detecting real life regimes' changes that are
difficult to see with conventional methods. Therefore, an additional potential use of the index is as an early indicator in economics for recessions and
market crashes [Piskun and Piskun (2011)], in seismology for earthquakes
[Zimatore *et al.* (2017)], etc.

An application to economics has been shown in [Orlando and Zimatore
(2017)] where we have run a RQE on USA GDP data versus recessions (see
Table A) as displayed in Fig. 35.

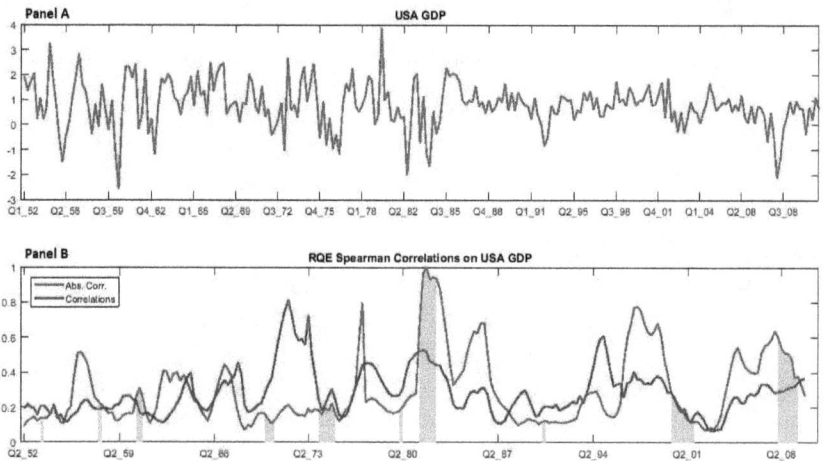

Figure 35   Maximum correlations (in blue) between RQE measures vs recession periods
(in grey). As shown in the figure a change in the index is often linked to a recession.
Spearman correlations (panel B) vs USA GDP (panel A). RQE absolute correlation
(in blue, panel B) is displayed next to correlation (red, panel B). See how the RQE
correlation calculated as in Eq. (12.1) is more reactive than the other and it is able to
detect more fine changes in the original times series. Difference in the x-axis numbering
between the picture above and below, is due to the windowing mechanism.

12.4.2   *RQE Correlation Index on a Kaldor-Kalecki Simulation*

As already mentioned in [Orlando (2018)], [Orlando (2016)] we described
a Kaldor-Kalecki model which is deterministic but displays random behaviour. For this analysis we have run a simulation of the variable income
(Y) and applied the RQA. After having defined a recession as a period in
which there are at least 3 consecutive negative variations of Y, we have
greyed those period in Fig. 36.

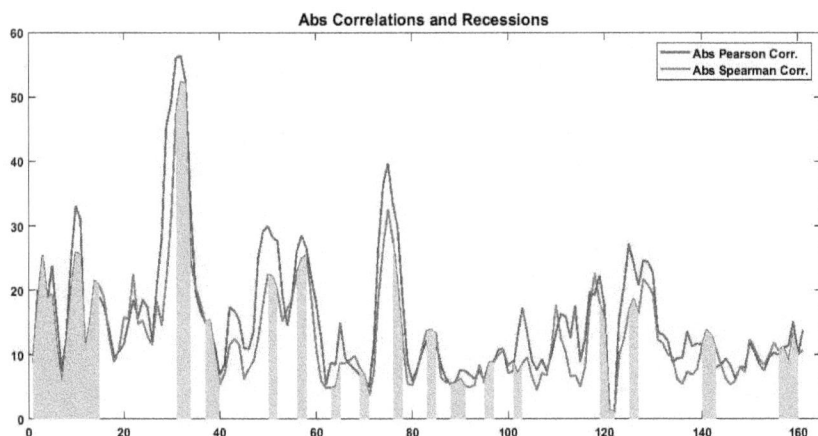

Figure 36 Absolute correlations between RQE measures vs recession periods (in grey) as obtained from a Kaldor-Kalecki model. As shown in the figure a significant change in the index is often linked to a recession. RQE Pearson correlation (in blue) is displayed next to Spearman correlation (red).

Once again the RQE correlation is more reactive than the other and it is able to detect more finely changes in the original times series.

12.5 Conclusions

RQA is meant to be an efficient and relatively simple tool in non-linear analysis. As the technique permits the identification of sudden phase-changes it may be suitable to study business cycles for, possibly, early detection of recessions.

This paper resumes the concept and results on the RQE correlation index in [Orlando and Zimatore (2017)] which was tested positively on a known signal. The application of the said index to USA GDP versus recessions has been shown as well. Finally the index was tested on the Y as generated from a Kaldor-Kalecki model giving similar results.

In future, through a new extensive data collection, RQA and PCA will be performed in order to assess the suitability of those techniques to studying business time series. Moreover RQA, PCA and statistical analysis will be applied on a Kaldor-Kalecki [Orlando (2016)], [Orlando (2018)] model to see whether it can generate series comparable to the real ones.

Acknowledgements

The authors are grateful to the referees and to their colleagues Carlo Lucheroni (School of Science and Technologies, University of Camerino) and Giovanni Taglialatela (Department of Economics and Finance, University of Bari). Special thanks go to Nicola Basile (Department of Mathematics, University of Bari) for his comments and helpful discussions.

Appendix

Table A   USA Recessions.

From		To	
Quarter	Year	Quarter	Year
Q4	1948	Q4	1949
Q3	1953	Q1	1954
Q4	1957	Q1	1958
Q3	1960	Q1	1961
Q1	1970	Q4	1970
Q1	1974	Q2	1975
Q1	1980	Q2	1980
Q3	1981	Q4	1982
Q3	1990	Q1	1991
Q2	2001	Q4	2001
Q1	2008	Q3	2009

US. Bureau of Economic Analysis [BEA (2016)]

Coupled crystal oscillator system and timing device

Antonio Palacios

San Diego State University, USA

13.1 Introduction

We present a computational and analytical study of a network-based model of a high-precision, inexpensive, Coupled Crystal Oscillator System and Timing (CCOST) device. A bifurcation analysis of the network dynamics shows a wide variety of collective patterns, mainly various forms of discrete rotating waves and synchronization patterns. Results from computer simulations seem to indicate that, among all patterns, the *standard* traveling wave pattern in which consecutive crystals oscillate out of phase by $2\pi/N$, where N is the network size, leads to phase drift error that decreases as $1/N$ as opposed to $1/\sqrt{N}$ for an uncoupled ensemble. The results should provide guidelines for future experiments, design and fabrication tasks.

13.2 Modeling

A crystal oscillator circuit sustains oscillation by taking a voltage signal from the quartz resonator, amplifying it, and feeding it back to the resonator. The frequency of the crystal is slightly adjustable by modifying the attached capacitances. A varactor, a diode with capacitance depending on applied voltage, is often used in voltage-controlled crystal oscillators, VCO. The analog port of the VCO chip is modeled by a nonlinear resistor R^-, which obeys the voltage-current relationship

$$v = -ai + bi^3,$$

where a and b are constant parameters. In addition, parasitic elements can be represented by a series resonator (L_2, C_2, R_2) connected in parallel with the nonlinear resistor. The resulting circuit, depicted in Fig. 37(left), forms a two-mode resonator model. see Fig. 37.

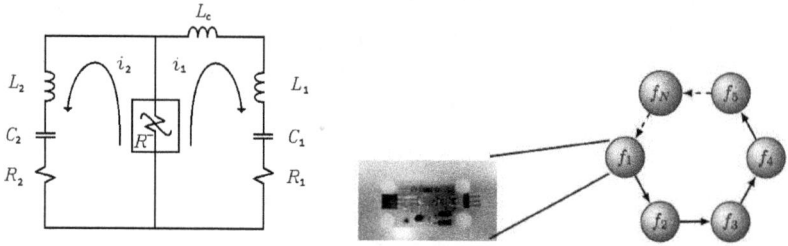

Figure 37   (Left) Two-mode crystal oscillator circuit. A second set of spurious RLC components (R_2, L_2, C_2) are introduced by parasitic elements. (Right) CCOST Concept.

Applying Kirchhoff's voltage law yields the following governing equations

$$L_j \frac{d^2 i_j}{dt^2} + R_j \frac{di_j}{dt} + \frac{1}{C_j} i_j = [a - 3b(i_1 + i_2)^2] \left[\frac{di_1}{dt} + \frac{di_2}{dt} \right], \qquad (13.1)$$

where $j = 1, 2$ and L_c has been included in L_1.

13.3   Governing Equations for Coupled System

In this section we consider a Coupled Crystal Oscillator System (CCOST) made up of N, assumed to be identical, crystal oscillators. Typical coupling topologies include unidirectional and bidirectional coupling in a ring fashion. Figure 37(right) shows the former case. The spatial symmetry of the unidirectionally coupled ring is described by the group Z_N of cyclic permutations of N objects. In the bidirectionally coupled case the symmetry group is D_N, which describes the symmetries of an N-gon.

Applying Kirchhoff's law to the CCOST network with unidirectional coupling yields the following (dimensionless version) governing equations

$$\frac{d^2 i_{k,1}}{dt^2} + \Omega_1^2 i_{k,1} =$$

$$\varepsilon \left\{ -R_1 \frac{di_{k,1}}{dt} + \left[a - 3b\big(i_{k,1} + i_{k,2} - \lambda\big[i_{k+1,1} + i_{k+1,2}\big]\big)^2 \right] \right.$$

$$\left. \left[\frac{di_{k,1}}{dt} + \frac{di_{k,2}}{dt} - \lambda \left(\frac{di_{k+1,1}}{dt} + \frac{di_{k+1,2}}{dt} \right) \right] \right\}$$

$$\frac{d^2 i_{k,2}}{dt^2} + \Omega_2^2 i_{k,2} =$$

$$\varepsilon L_r \left\{ -R_2 \frac{di_{k,2}}{dt} + \left[a - 3b\big(i_{k,1} + i_{k,2} - \lambda\big[i_{k+1,1} + i_{k+1,2}\big]\big)^2 \right] \right.$$

$$\left. \left[\frac{di_{k,1}}{dt} + \frac{di_{k,2}}{dt} - \lambda \left(\frac{di_{k+1,1}}{dt} + \frac{di_{k+1,2}}{dt} \right) \right] \right\},$$

(13.2)

where $L_{k,1} = L_1$, $L_{k,2} = L_2$, $R_{k,1} = R_1$, $R_{k,2} = R_2$, $C_{k,1} = C_1$ and $C_{k,2} = C_2$. Letting $t = \sqrt{L_1 C_1}\tau$, $\Omega_1^2 = 1$, $\Omega_2^2 = \frac{L_1}{L_2}\frac{C_1}{C_2}$, $L_r = \frac{L_1}{L_2}$, $\varepsilon = \sqrt{\frac{C_1}{L_1}}$. The new time variable τ has been relabeled as t.

13.4 Averaging

After applying the following set of invertible coordinates transformations

$$i_{kj} = x_{kj} \cos\phi_{kj};$$
$$i'_{kj} = -\Omega_j x_{kj} \sin\phi_{kj};$$
$$i''_{kj} = -\Omega_j x'_{kj} \sin\phi_{kj} - \Omega_j^2 x_{kj} \cos\phi_{kj} - \Omega_j x_{kj}\psi'_{kj} \cos\phi_{kj};$$
$$\phi_{kj} = \Omega_j t + \psi_{kj};$$

(13.3)

for $j = 1, 2$ we arrive at the following set of equations, written symbolically as:

$$\begin{bmatrix} \mathbf{x}'_k \\ \phi'_k \\ \phi'_s \end{bmatrix} = \begin{bmatrix} 0 \\ 0 \\ \mathbf{\Omega}^0 \end{bmatrix} + \varepsilon \begin{bmatrix} \mathbf{X}^{[1]}(\mathbf{x}_k, \phi_k + \phi_s, \phi_{k+1} + \phi_s, \varepsilon) \\ \mathbf{\Omega}^{[1]}(\mathbf{x}_k, \phi_k + \phi_s, \phi_{k+1} + \phi_s, \varepsilon) \\ 0, \end{bmatrix}.$$

(13.4)

where $\mathbf{x}_k = (x_{k1}, x_{k2})$, $\phi_k = (\phi_{k1}, \phi_{k2})$ and $\mathbf{\Omega}^0 = (\Omega_1, \Omega_2)$. These equations include the shift $\phi_k \mapsto \phi_k + \phi_s$ and $\phi_{k+1} \mapsto \phi_{k+1} + \phi_s$, where $\phi_s = (\phi_{s1}, \phi_{s2})$.

After applying the averaging method, we arrive at a new set of equations, which can be written in complex form to facilitate analysis. The equations are of the form

$$\dot{z}_{k1} = f_1(z_{k1}, z_{k2}, z_{k+1,1}, z_{k+1,2}, \mu)$$
$$\dot{z}_{k2} = f_2(z_{k1}, z_{k2}, z_{k+1,1}, z_{k+1,2}, \mu),$$

(13.5)

where μ is a vector of parameters. A similar set of equations are obtained for the bidirectional case. The complete equations can be found in [Buono et al. (2018a,b)]. The symmetry of these averaged amplitude-phase equations is captured by the groups $Z_N \times \mathbf{O(2)} \times \mathbf{O(2)}$ and $\mathbf{D}_N \times \mathbf{O(2)} \times \mathbf{O(2)}$ for the unidirectional and bidirectional coupling cases, respectively. A complete analysis of the equations can be found in [Buono et al. (2018a,b)]. We summarize the main results. Steady-states of the averaged system with symmetry group $\Sigma \subset \Gamma \times \mathbf{SO(2)}$, with $\Gamma = Z_N$ and $\Gamma = \mathbf{D}_N$ lead to periodic solutions with spatio-temporal symmetry $\Sigma \subset \Gamma \times \mathbf{S}^1$. Then, the tangent space to the trivial steady-state can be decomposed along irreducible representations of the Z_N and $\mathbf{D}_N$ actions and thus we obtain a block diagonalization of the linearization of the complexified governing equations. Symmetry-preserving and symmetry-breaking bifurcations are then determined by examining the eigenvalues computed directly from the block diagonalization. Criticality computations are also performed to determine the direction of bifurcations.

13.5   Phase Drift

Phase error is defined as the drift of the period of oscillation of an oscillating system away from the expected period length. To study phase drift, the governing equations are rewritten in Langevin form

$$d_t X_k = F(X_k) - \lambda \sum_{j \to k}^{N} h(X_j, X_k) + \eta_k$$

$$d_t \eta_k = -\frac{\eta_k}{\tau_c} + \frac{\sqrt{2D}}{\tau_c} \xi_k,$$

(13.6)

where the noise function η_k is assumed to be Gaussian, band-limited, having a zero mean, a variance σ^2, and have a specific correlation time, τ_c. The noise is assumed to not drive the dynamics of the system, this corresponds to $\tau_f \ll \tau_c$, where τ_f is the time-constant of each oscillator [Gardiner (2003); Wio et al. (2012)]. $X_k = [i_{k1}, i'_{k1}, i_{k2}, i'_{k2}]$ is the state variable of each crystal oscillator, τ_c, D are correlation time and noise intensity respectively, F represents the internal dynamics of each oscillating unit, i.e., each crystal oscillator, h is the coupling function between two oscillators, in which the summation is taken over those cells j that are coupled to each cell k, λ is the coupling strength, ξ_k is a Gaussian distributed random variable with zero mean, and standard deviation σ.

Figure 38(top-left) illustrates the performance with respect to the scaling exponent, i.e., this figure is a log plot phase error, $Err(N, \lambda) = N^{m(\lambda)}$. Samples are taken for 100 values of λ. For each value of λ, the mean phase error for 50 repeated simulations is calculated for $N = 3, 5, \ldots, 21$. Then a least squares regression is performed on the log of these values, producing the scaling exponents depicted in Fig. 38. This analysis suggests that strong coupling is preferable to weak coupling to produce optimal scaling. From Fig. 38, the optimal scaling is found at $\lambda = 0.99$ with $m = -0.8947$. Figure 38(top-right) illustrates the design and network response captured by an oscilloscope. The white box in the figure contains appropriate potentiometers to control the gain of the operational amplifiers, which in turn, are used to manipulate coupling strength, and thus, control the network response to the desired pattern of oscillation.

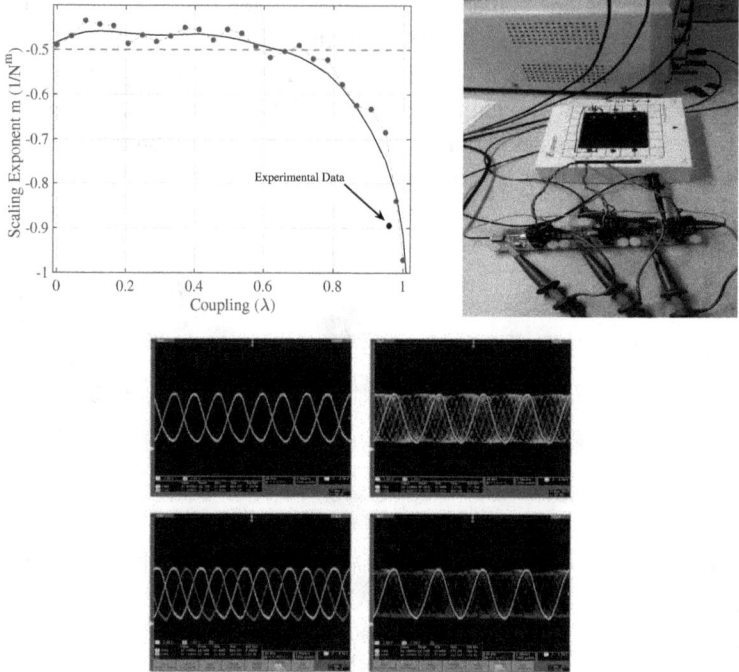

Figure 38 (Top) Experimental realization of a network of coupled crystal oscillators implemented via PIC boards. (Bottom-left) Experimental measurements for $N = 2$ and $N = 3$ reveal, as expected, a traveling wave pattern among the oscillations. (Bottom-right) When the oscillators are uncoupled the pattern disappears.

Chaotic behaviors in a system with stable equilibrium

Viet-Thanh Pham[1,*], Sajad Jafari[2], Christos Volos[3] and Luigi Fortuna[4]

[1] *Modeling Evolutionary Algorithms Simulation and Artificial Intelligence,
Faculty of Electrical & Electronics Engineering, Ton Duc Thang
University, Ho Chi Minh City, Vietnam*
[2] *Biomedical Engineering Department, Amirkabir University of
Technology, Tehran, Iran*
[3] *Laboratory of Nonlinear Systems, Circuits & Complexity (LaNSCom),
Department of Physics, Aristotle University of Thessaloniki, Thessaloniki,
Greece*
[4] *Dipartimento di Ingegneria Elettrica Elettronica e Informatica,
Universita degli Studi di Catania, Catania, Italy*

14.1 Introduction

A considerable amount of literature has been published on chaotic systems [Lorenz (1963); Rössler (1976); Chen and Ueta (1999); Fortuna et al. (2009); Sprott (2010)]. However, these studies focused particularly on chaotic systems with the presences of unstable equilibrium points. Recently, there has been renewed interest in systems with stable equilibrium points, especially systems with only one stable equilibrium [Wang and Chen (2012); Molaie et al. (2013)]. Such systems relate to a class of special systems, in which their attractors are "hidden" from the computing viewpoint [Leonov and Kuznetsov (2013); Dudkowski et al. (2016)]. A novel three-dimensional system with stable equilibrium is proposed and investigated in this work.

14.2   The System and its Equilibrium

Motivated by published works [Wang and Chen (2012); Molaie *et al.* (2013)], a new three-dimensional system is considered:

$$\begin{cases} \dot{x} = z, \\ \dot{y} = -x - z, \\ \dot{z} = -ax + by - z + cx^2 + z^2 + xy, \end{cases} \tag{14.1}$$

where a, b, c are positive parameters.

By setting

$$\begin{cases} z = 0, \\ -x - z = 0, \\ -ax + by - z + cx^2 + z^2 + xy = 0, \end{cases} \tag{14.2}$$

we find that the system (14.1) has only one equilibrium point $E(0,0,0)$. The Jacobian matrix of the system at the equilibrium E is

$$J_E = \begin{bmatrix} 0 & 0 & 1 \\ -1 & 0 & -1 \\ -a & b & -1 \end{bmatrix}. \tag{14.3}$$

Therefore, we get the following characteristic equation of the system:

$$A_3\lambda^3 + A_2\lambda^2 + A_1\lambda + A_0 = 0, \tag{14.4}$$

with

$$\begin{cases} A_3 = A_2 = 1, \\ A_1 = a + b, \\ A_0 = b. \end{cases} \tag{14.5}$$

It is trivial to confirm that

$$\begin{cases} A_3, A_2, A_1, A_0 > 0, \\ A_2 A_1 > A_3 A_0, \end{cases} \tag{14.6}$$

for the positive parameters $a, b, c > 0$. According to the Routh-Hurwitz stability criterion, there is one stable equilibrium in the system.

Interestingly, chaos is observed in the system with stable equilibrium for $a = 0.9$, $b = 3$, $c = 0.8$ and $(x(0), y(0), z(0)) = (0, -1, 3)$ as shown in Fig. 39.

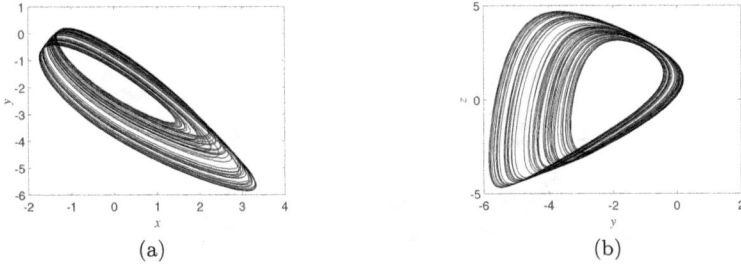

(a)   (b)

Figure 39   Chaos in the system with stable equilibrium. (a) $x - y$ plane. (b) $y - z$ plane.

14.3   Dynamics of the System

Focusing on the case where there is the presence of the stable equilibrium, we have discovered the dynamics of the system (14.1) by changing the positive parameter a for $b = 3$ and $c = 0.8$. As can be seen from the bifurcation diagram (Fig. 40) and the diagram of maximal Lyapunov exponents (Fig. 41), the system (14.1) displays chaotic and periodic dynamics. In addition, a route to chaos through the mechanism of period doubling is observed. Chaos dynamics can be found for $a \in (0.9, 0.9082)$.

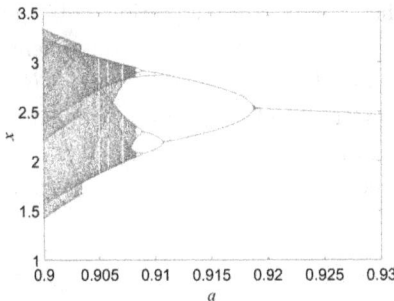

Figure 40   Bifurcation diagram of the system (14.1), which has only one stable equilibrium $E(0, 0, 0)$.

14.4   Circuit Design

We have designed an electronic circuit for the system (14.1) as presented in Fig. 42. The circuit includes five operational amplifiers $(U_1 - U_5)$, thirteen resistors, three capacitors and three analog multipliers $(U_6 - U_8)$. By

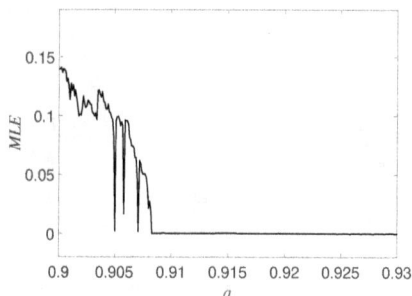

Figure 41   Calculated maximal Lyapunov exponents of the system (14.1) for $b = 3$ and $c = 0.8$.

denoting the voltages at the outputs of operational amplifiers U_1, U_2, U_3 as X, Y, Z, the circuit's equation has the form:

$$\begin{cases} \dot{X} = \frac{1}{R_1 C_1} Z, \\ \dot{Y} = -\frac{1}{R_2 C_2} X - \frac{1}{R_3 C_2} Z, \\ \dot{Z} = -\frac{1}{R_4 C_3} X + \frac{1}{R_5 C_3} Y - \frac{1}{R_6 C_3} Z + \frac{1}{R_7 C_3 10V} X^2 \\ \quad + \frac{1}{R_8 C_3 10V} Z^2 + \frac{1}{R_9 C_3 10V} XY. \end{cases} \quad (14.7)$$

It is trivial to see that the circuit's equation (14.7) emulates the system with stable equilibrium (14.1). Values of components in the circuit are $C_1 = C_2 = C_3 = 10$ nF, $R_1 = R_2 = R_3 = R_6 = R = 90$ kΩ, $R_4 = 100$ kΩ, $R_5 = 30$ kΩ, $R_7 = 11.25$ kΩ, and $R_8 = R_9 = 9$ kΩ. The circuit has been implemented in PSpice and PSpice attractors are reported in Fig. 43. Obviously, the designed circuit generates chaotic signals.

14.5   Conclusions

A system with stable equilibrium has been studied in our work. The equilibrium of the system is calculated and its stability is analyzed. It is interesting that there is coexistence of a chaotic attractor and a stable equilibrium. In addition, we have designed an electronic circuit based on the theoretical system. Chaotic signals have been observed in the implemented circuit. In future investigations, it might be possible to use such a system for developing chaos-based applications.

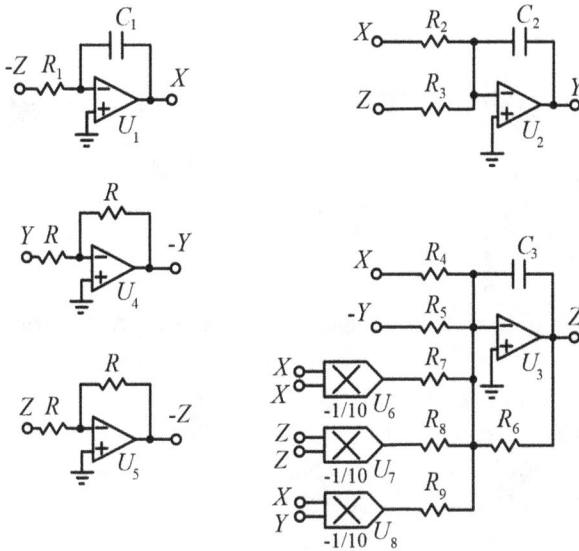

Figure 42 Circuit designed for the system by using operational amplifier–based approach.

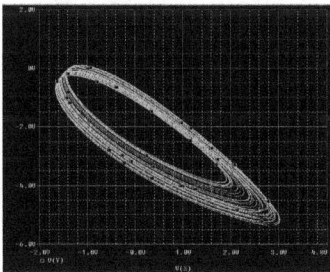

(a)

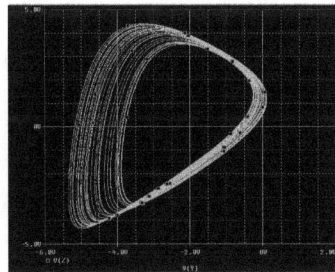

(b)

Figure 43 Chaos in the circuit, which is obtained by PSpice. (a) $X - Y$ plane. (b) $Y - Z$ plane.

Portraying human motor imagery: A MEG study

P. Chholak[1], V. A. Maksimenko[2], N. S. Frolov[2], A. E. Hramov[2] and A.N. Pisarchik[1,2]

[1] *Center for Biomedical Technology, Technical University of Madrid, Spain*
[2] *Research and Educational Center 'Artificial Intelligence Systems and Neurotechnology', Yuri Gagarin State Technical University of Saratov, Russia*

15.1 Introduction

Mental imagination of movements referred to as *motor imagery* (MI) [Jeannerod (1994)] manifests as a result of the rehearsal of a given motor act in the working memory without any overt movement of the corresponding muscle. It is classified into two categories, namely, visual imagery (VI) and kinesthetic imagery (KI). VI consists of visualization of the subject moving a limb, that does not require any special training or sensing of the muscles. In contrast, KI is the feeling of muscle movement, that is often realized by athletes or specially trained persons [Mizuguchi *et al.* (2012)].

Previous studies using functional magnetic resonance imaging (fMRI) [Solodkin *et al.* (2004); Hanakawa *et al.* (2008)] indicate that brain activity associated with KI is similar to real movement because it includes control of muscle contractions which are then blocked at some level of the motor system by inhibitory mechanisms. This enables MI to share a part of the same neuronal network which is involved in real movement. Diverse transcranial magnetic stimulation (TMS) studies [Izumi *et al.* (1995); Kasai *et al.* (1997); Stinear and Byblow (2003); Liang *et al.* (2007); Mizuguchi *et al.* (2012)] also confirm this.

15.2 Methods

The neurophysiological data were acquired with the Vectorview MEG system (Elekta AB). Fastrak digitizers (Polhemus) were used to obtain the three-dimensional head shape. The experimental study consisted of eight right-handed untrained voluntaries, six males and two females. The subjects were sat in a comfortable reclining chair with their legs straight and arms resting on an armrest in front of them, as shown in Fig. 44. The whole experiment was divided into four series with one-fourth of the total number of trials in each series. Each series consisted of trials randomly chosen for each of the four limbs (left/right arm/leg imagery). The subject was informed on the screen in front of him/her about which limb he/she must imagine to move after hearing a beep. The imaginary movement of each limb counted as one trial.

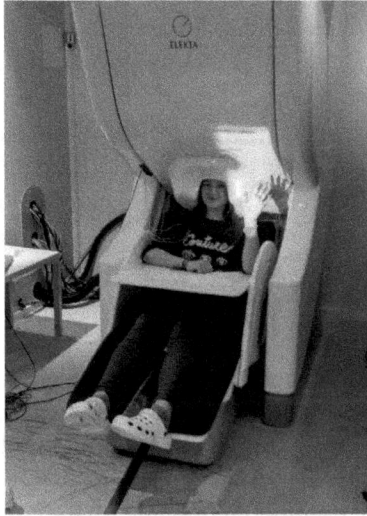

Figure 44   Overview of the MEG machine with a subject.

Artifacts in the MEG recordings due to eye movement, heartbeat, breathing, and blinking were removed using the temporal signal-space separation method of Taulu and Hari [Taulu and Hari (2009)] and other techniques. Once the events were marked at the beginning of each limb's MI from a log file, we extracted 5-s trials just after these marks. Similarly, 20-s trials corresponding to the resting state with closed eyes were also marked and extracted as the background activity of each subject.

The time-frequency structure of MEG signals was analysed with the help of a wavelet-based approach. For each limb, we used the Morlet wavelet to evaluate the time-frequency spectrogram (TFS) for all extracted epochs, and then averaged the TFSs for every limb. Then, the TFS was also averaged over a desired frequency range, such as δ (1–5 Hz), α (8–12 Hz), β (15–30 Hz) and μ (8–30 Hz). The same process was repeated over the background resting state using the same parameters. To evaluate event-related synchronization/desynchronization (ERS/ERD), we took a difference between the averaged-over-time spectrograms of the trials and the background and then normalized it to the background spectrogram. This normalized difference was positive for ERS and negative for ERD.

For classification of the brain states associated with MI, we used a popular type of artificial neural networks (ANN) called multilayer perceptron (MLP) [Haykin (2009)]. We constructed the MLP which consisted of an input layer with selected number of MEG channels for training/testing the network, followed by three hidden layers with 30, 15 and 5 neurons. The output layer comprised of a single neuron. We used the training algorithm called *scaled conjugate gradient* because it provides higher efficiency for pattern recognition problems than other algorithms, such as the Levenberg-Marquardt. First, we trained the ANN using 75% of MEG trials and then tested it with the rest of the 25% trials. To improve accuracy, we only used 2-s trials for all subjects chosen within 1.5–3.5 s intervals of each 5-s epoch. Trials shorter than 2-s trials did not provide sufficient information.

15.3   Results

The time series of wavelet energy is unique for each limb and each subject. The TFS averaged over the frequency band and limb trials (Fig. 45) revealed synchronous behavior of alpha and beta activity for each limb. The signatures of a similar trend in delta band are also seen in the figure.

To get further insight, we selectively picked MEG data for ANN training/testing in classifying left and right hand MI in one of the KI subjects and compared the resulting performance to evaluate which data was more relevant to the MI content.

We chose 15 channels in the most significant ERS/ERD brain area, the inferior parietal lobe (IPL). IPL is known to play a significant role in KI-dominated MI inhibitory control ([Lebon *et al.* (2012)]). 102 channels represented magnetometers, whereas 72 channels represented the active contributors to the principal components of the MEG trials. We found that

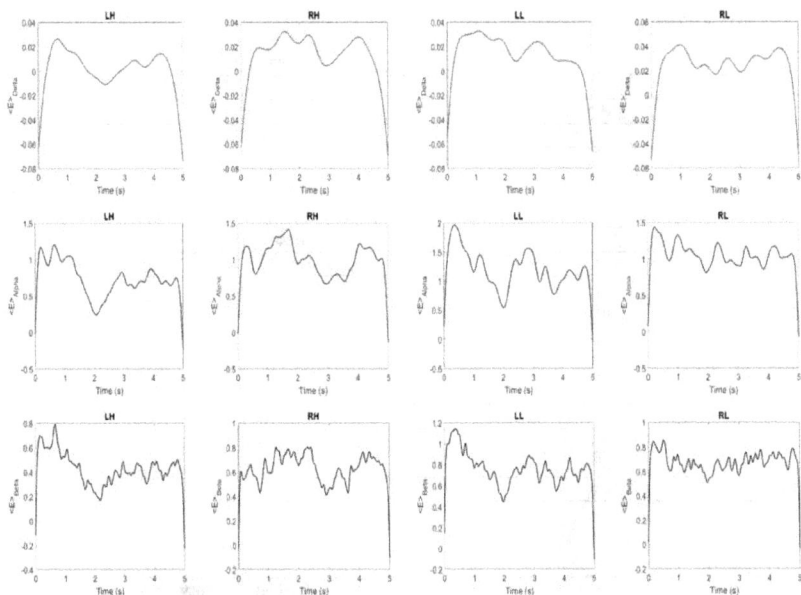

Figure 45 Wavelet energy time series for delta (magenta), alpha (blue) and beta (black) bands for one of the subjects. Each column from left to right represents MI for left hand (LH), right hand (RH), left leg (LL) and right leg (RL), respectively.

the principal component analysis (PCA) gave the best performance in our study, balancing between too-much and too-less information (Fig. 46(a)). However, we should note that 12 from the 15 channels belonged to the best performing 72 channels. We pre-processed the MEG time-series before training/testing using a low-pass filter. By changing the cut-off frequency, we obtained the best accuracy when we included the μ-band (Fig. 46(b)).

15.4   Conclusion

The synchronous behavior hints towards the possibility of μ waves, that are specific to movement related tasks ([Yin *et al.* (2016)]), being carried over δ waves acting as a general carrier. Machine learning studies further highlight that the content of motor imagery, being used to choose between hands, is communicated through μ waves, because the ANN performance drastically improves after inclusion of this frequency band. The improvement in the ANN performance using channels in the vicinity of the inferior parietal lobe re-emphasizes its role in MI.

X = 72; Y = 30 Hz

	% Mean accuracy
Mean	77.5
Min	75
Max	83.33

X = 15; Y = 30 Hz

	% Mean accuracy
Mean	28.33
Min	8.33
Max	41.67

X = 102; Y = 30 Hz

	% Mean accuracy
Mean	63.33
Min	41.67
Max	75

(a)

X = 72; Y = 30 Hz

	% Mean accuracy
Mean	**77.5**
Min	**75**
Max	**83.33**

X = 72; Y = 10 Hz

	% Mean accuracy
Mean	65.83
Min	58.33
Max	75

X = 72; Y = 4 Hz

	% Mean accuracy
Mean	36.67
Min	25
Max	50

(b)

Figure 46   ANN performance accuracy for classification between left and right hands MI in one of the KI subjects involving the variation of (a) input MEG channels (X) and (b) cut-off frequency (Y).

Acknowledgements

This work was supported by the Russian Foundation for Basic Research (Grant 16-29-08221) in the part of data analysis. A. N. P. acknowledges support for the experimental research from the Ministry of Economy and Competitiveness (Spain) (project SAF2016-80240). A. E. H. acknowledges individual support from the Ministry of Education and Science (Russia) (Agreement No. 3.4593.2017/6.7) for the development of the experiment design.

Study of the behavior of a chaotic oscillator with memory elements

D. A. Prousalis[1], C. K. Volos[1], I. N. Stouboulos[1] and I. M. Kyprianidis[1]

[1]*Laboratory of Nonlinear Systems - Circuits & Complexity (LaNSCom), Department of Physics, Aristotle University of Thessaloniki, Greece*

16.1 Introduction

Chua introduced the missing fourth circuit element, the memristor, in 1971 [Chua (1971)] according to the theory of electrical circuits combining the charge q and the flux ϕ. A general concept of memristive systems was developed by [Chua and Kang (1976)]. A solid-state implementation of memristors was reported by Strukov *et al.* (2008) at Hewlett-Packard Laboratories. [Ventra *et al.* (2009)] proposed other "mem-elements", such as memcapacitor and meminductor. Research on memristor-based chaotic systems becomes a focal research topic in both the technological and the application domain [Wu *et al.* (2011); Volos *et al.* (2011); Yang *et al.* (2013); Driscoll *et al.* (2010)]. Also, the design of memristor-based chaotic oscillators has been developed by replacing the nonlinear part of chaotic dynamical systems with memory elements [Chua (2011); Itoh and Chua (2008)]. In order to explore the circuit characteristics of memristor, memcapacitor, and meminductor, some chaotic oscillators based on these memory elements were designed, and some special phenomena were found, such as coexisting attractors, hidden attractors, and extreme multistability [Kuznetsov *et al.* (2010); Bao *et al.* (2017)].

In this work, a chaotic oscillator with memory elements is presented. It is found that the proposed oscillator can exhibit some complex phenomena, such as chaos and coexisting attractors. Furthermore, many basic dynamical behaviors of the including equilibrium sets with various circuit parameters, are also investigated numerically. Our analysis results show that the proposed system possesses complex dynamics. The system is solved numerically by applying the fourth-order Runge-Kutta algorithm and various tools of nonlinear dynamics such as the phase portraits have been used.

This research work is organized as follows. In Sec. 16.2 are presented the chaotic circuit based on memcapacitor and memristor with the model of the memcapacitor and memristor. In Sec. 16.3 the simulation results of the system's behavior is presented. Section 16.4 concludes this work with a summary of the main results.

16.2 Chaotic Circuit Based on Memcapacitor and Memristor

16.2.1 *The devices of Memristor and Memcapacitor*

In the present work the TiO_2 memristor model has been used, with the following form of the Memristance:

$$M = R_{on}\frac{w}{D} + R_{off}(1 - \frac{w}{D}), \quad \frac{dw}{dt} = \mu_\nu \frac{R_{on}}{D} i(t) \tag{16.1}$$

where M is called memristance; R_{on} and R_{off} are the low resistance and high resistance for $w = D$ and $w = 0$, respectively; μ_ν is the dopant mobility; $i(t)$ is the current through the memristor; w is the thickness of TiO_{2-x} layer and D represents the total thickness of the memristor. The form of $v(t)$ is:

$$v(t) = (a - b \int_{-\infty}^{t} i(\tau)d\tau)i(t) \tag{16.2}$$

where $a = R_{off}$ and $b = (R_{off}R_{on})\mu_V R_{on}/D^2$. Obviously, the proposed memristive system exhibits a pinched hysteresis loop in the input-output plane.

In 2009, Di Ventra extended the concept of memory devices from memristor to memcapacitor and meminductor.

Analogous to the aforementioned, we obtained the simplified model of a memcapacitor with the following form of the device:

$$C_M^{-1}(x, q, t)q(t) \\ \frac{dx}{dt} = f(x, q, t). \tag{16.3}$$

The form of $V(t)$ of the memcapacitor is:

$$V(t) = (c - d \int_{t_0}^{t} q(\tau)d\tau)q(t) \tag{16.4}$$

where $c - d \int_{t_0}^{t}(q\tau)$ is the inverse memcapacitance C_m^{-1}.

16.2.2 The System

Wang *et al.* (2016) designed a nonlinear circuit, as shown in Fig. 47, which contains an active memcapacitor C_m, a memristor M, a inductor L, a capacitor C, and a resistor R:

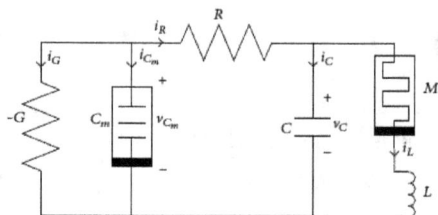

Figure 47 Nonlinear circuit with a memristor and a memcapacitor.

The state equations of the proposed circuit can be obtained by Kirchhoff's current and voltage laws:

$$C\frac{di_L}{dt} = -i_L + \frac{1}{R}(v_{C_m} - v_c)$$

$$L\frac{dv_C}{dt} = v_c - Mi_L$$

$$\frac{dq_{C_m}}{dt} = Gv_{C_m} + \frac{1}{R}(v_{C_m} - v_c)$$

$$\frac{dq_M}{dt} = i_L$$

(16.5)

where $v_{C_m} = (c - d\sigma_{C_m})q_{C_m}$, $M = a - bq_M$ and $\sigma_{C_m} = \int q_{C_m}dt$, called integration variable of charge.

Normilizing the equations:

$$\frac{dx}{dt} = n(j((c - dw))z - x) - y,$$

$$\frac{dy}{dt} = m(x - (a - bv)y),$$

$$\frac{dz}{dt} = k(c - dw)z + j(x - (c - dw))z,$$

(16.6)

$$\frac{dw}{dt} = z,$$

$$\frac{dv}{dt} = y$$

where $x = v_c$, $y = i_L$, $z = q_{c_m}$, $w = \sigma_{C_m}$, $v = q_M$, $m = 1/L$, $n = 1/C$, $j = 1/R$, $k = G$, $\sigma_{C_m} = \int q_{C_m}dt$.

16.3 Analysis of the System

Firstly we started to study the system from dynamical point of view. Some special phenomena were found, such as coexisting attractors, hidden attractors, and extreme multistability.

By letting $\dot{x} = \dot{y} = \dot{z} = \dot{w} = \dot{v} = 0$ the system for every set of values of the parameters has number of equilibria $E(0, 0, 0, w, v)$, where v and w are arbitrary real numbers.

The Jacobian of the system is:

$$\mathbf{J} = \begin{pmatrix} -jn & -n & jn(g - dw) & -djnz & 0 \\ m & m(-a + bv) & 0 & 0 & bmy \\ j & 0 & k(g - dw) + j(-g + dw) & djz - dkz & 0 \\ 0 & 0 & 1 & 0 & 0 \\ 0 & 1 & 0 & 0 & 0 \end{pmatrix}$$

The characteristic equation $M = det(A - xI) = 0$ of equilibrium point set as for $x = y = z = 0$ and for $a = 0.01$, $m = 7.35$, $n = 0.17$, $j = 4.8$, $k = 2.1$, $b = 0.1$, $g = 0.7$, $d = -2$ is:

$$M = \lambda^2 (j^2 n(\lambda + m(a - bv))(g - dw) - (\lambda^2 + a\lambda m + j\lambda n + mn + ajmn - bm(\lambda + jn)v)(\lambda + (j - k)(g - dw))) = 0$$

The stability of equilibria of a dynamical system is determined by the sign of real parts of eigenvalues of the Jacobian matrix. As it is clear the $\lambda_1 = \lambda_2 = 0$ as expected because the system is degenerative. The rest eigenvalues λ_3, λ_4, λ_5 of the Jacobian Matrix depend on the variables w, v. So, it is difficult to determine the stability of the equilibrium points. In Figs. 48, 49 the real parts of the eigenvalues of the Jacobian matrix are depicted. As it is shown the behavior of the equilibrium points is different for different w, v. So there are regions of equilibrium points which are stable multiplicities and regions where the equilibrium points are unstable multiplicities.

In previous it is proved that the system (16.6) has infinite number of equilibrium points. By changing the initial conditions of the variables x, y, z, w, v the system appears different behaviour. In phase portraits of Figs. 50, 51 the dependence on the initial conditions of the variables x, y, z, w, v and the co-existence of chaotic with various periodical states are presented.

16.4 Conclusion

The study of an oscillator with memory elements has been presented in this work. The system has rich dynamical behavior as confirmed by the example of the reported attractors as well as by the system's numerical study. The system shows very interesting phenomena such as such as co-existing attractors, hidden attractors, and extreme multistability. Due to

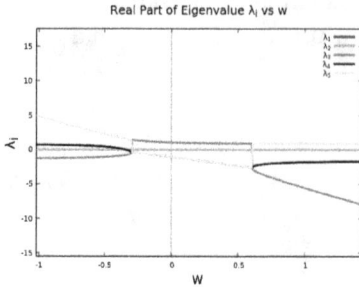

Figure 48   The dependence of the types and signs of the eigenvalues on the variable w for the plane $v = 0$.

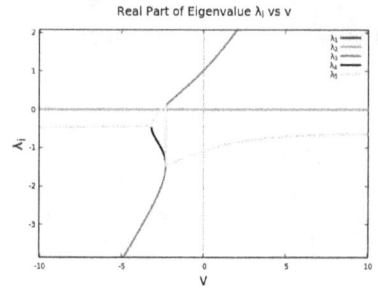

Figure 49   The dependence of the types and signs of the eigenvalues on the variable v for the plane $w = 0$.

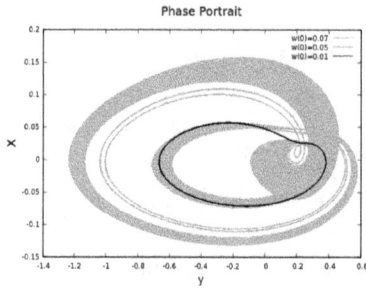

Figure 50   Different phase portraits where the blue line is for $w(0) = 0.07$, the black line is for $w(0) = 0.01$ and the green line is for $w(0) = 0.05$.

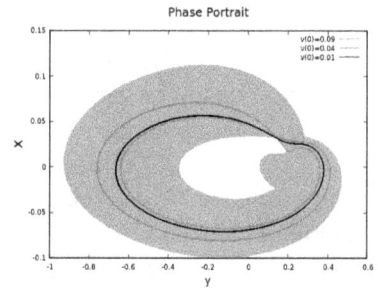

Figure 51   Different phase portraits where the blue line is for $v(0) = 0.09$, the black line is for $v(0) = 0.01$ and the purple line is for $v(0) = 0.04$.

the little knowledge about the special features of such system, future works will continue focusing on studying, from mathematical point of view, the properties of these dynamical systems, the case of coupling scheme of these oscillators and the circuit implementation of the coupled system.

Power amplification of an erbium-doped fiber laser array using multistability

J.O. Esqueda-de-la-Torre[1], R. Jaimes-Reátegui[1], J.H. García-López[1],
G. Huerta-Cuellar[1], Carlos E. Castañeda[1], C.E. Rivera-Orozco[1] and
A.N. Pisarchik[2]

[1] *Centro Universitario de los Lagos, Universidad de Guadalajara, Mexico*
[2] *Center for Biomedical Technology, Technical University of Madrid, Spain*

17.1 Introduction

In past decades, revolutionary progress has been achieved in research and commercialization of erbium-doped fiber lasers (EDFLs). The advantage of these lasers is the long interaction length of pumping light with active ions, which leads to a high gain and a single-transversal-mode operation produced by a suitable choice of the fiber core diameter and index step. These features make fiber lasers unique in communications, reflectometry, sensing, and medicine [Digonnet (2001); Luo and Chu (1998)]. The EDFLs with external modulation are nonautonomous systems in which polarization is adiabatically eliminated and the dynamics can be ruled by two rate equations, laser intensity and population inversion.

The dynamics of the EDFL exhibits the coexistence of different periodic and chaotic attractors (generalized multistability). In such systems, a particular state is determined by initial conditions [Saucedo-Solorio *et al.* (2003)]. The coexistence of multiple periodic attractors was reported theoretically and experimentally in EDFLs subject to loss [Saucedo-Solorio *et al.* (2003)] or pump modulation [Pisarchik *et al.* (2003a,b)]. The EDFLs are quite sensitive to any external perturbation that may destabilize their normal operation. Therefore, the knowledge of the dynamic behavior of these lasers under external modulation is of great importance for many applications [Pisarchik *et al.* (2005)]. Different conditions for the development of a chaotic motion have been found in EDFLs with harmonic modulation. First, a period-doubling route to chaos was observed in a bipolarized two-mode EDFL with harmonic pump modulation [Lacot *et al.* (1994)]. The authors also developed a model based on two coherently pumped coupled lasers. A quasi-periodic route to chaos was found in a dual-wavelength EDFL [Sanchez *et al.* (1995)]. Eventually, Luo *et al.* [Luo *et al.* (1998b)] revealed the coexistence of period-doubling and intermittency routes to chaos in a pump-modulated ring EDFL. They also reported on bistability (the coexistence of two periodic attractors) in this laser [Luo *et al.* (1998b,a)]. Synchronization of multistable EDFLs can be beneficial for industrial applications which require a high peak power [Boccaletti *et al.* (2018b)]. This paper is structured as follows. In Sec. 17.1, the main background and concepts are briefly explained. Then, in Sec. 17.2, the mathematical model and basics of this work are present. The instrumental design and results are described in Sec. 17.3. Finally, main conclusions are given in Sec. 17.4.

17.2   Model Description

The mathematical model of the normalized EDFL is given as follows [Reategui (2005)]

$$\frac{dx}{dt} = axy - bx + c(y + 0.3075),$$
$$\frac{dy}{dt} = -dxy - (y + 0.3075) + P_{pump}(1 - e^{-18\frac{1-(y+0.3075)}{0.6150}}),$$
(17.1)

where the variables x and y are proportional to the laser intensity and population inversion, respectively, and the laser parameters are $a = 6.6206 \times 10^7$, $b = 7.4151 \times 10^6$, $c = 0.0163$, and $d = 4.0763 * 10^3$. Since EDFL belongs to class-B lasers, an additional degree of freedom is required to observe nonlinear effects. Therefore, the pump modulation is added as $P_{pump} = 506(1 + m\sin(2\pi F_m t))$, where m and F are the modulation amplitude and frequency.

    The bifurcation diagram of local maxima of laser intensity x is shown in Fig. 52(a) as a function of the modulation frequency $1 < F_m < 100$ (kHz). In this figure, the multistability region is fixed to $F_m = 80$ KHz and $m = 1$, where four labeled branches are distinguished as period one (P1), period three (P3), period four (P4), and period five (5), i.e., we observe the coexistence of multistable attractors. The times series in Fig. 52(b)

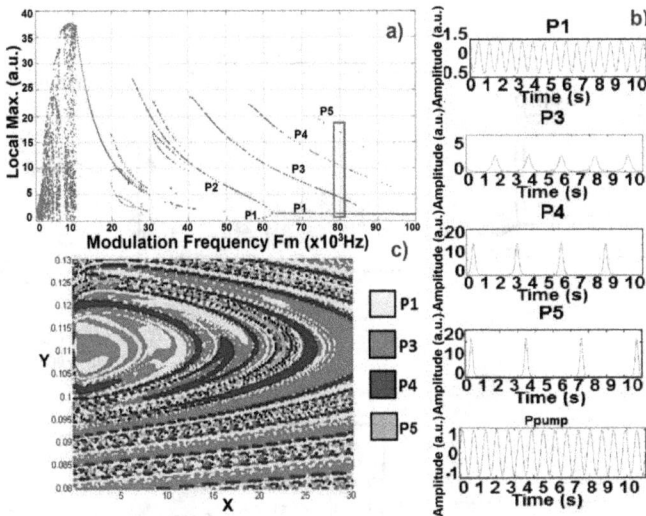

Figure 52 (a) Bifurcation diagram of pump-modulated EDFL, (b) times series illustrating the coexistence of P1, P3, P4, P5 attractors and pump modulation signal P_{pump}, and (c) basin of attraction of the coexisting periodic orbits: yellow – P1, red – P3, blue – P4 and green – P5.

illustrate these coexistent attractors which can be compared with the pump modulation signal shown in the lower panel. The periodic attractors P1, P3, P4 and P5 appear at subharmonic frequencies of the modulation frequency P_{pump}: $\frac{P1}{P_{pump}} = 1$, $\frac{P3}{P_{pump}} = 3$, $\frac{P4}{P_{pump}} = 4$ and $\frac{P5}{P_{pump}} = 5$, where the P5 attractor has the largest peak intensity. Figure 52(c) shows the basin of attraction of the coexisting periodic orbits. The yellow, red, blue and green colors represent the basins of P1, P3, P4 and P5 attractors, respectively, found for different initial conditions (x_0, y_0).

17.3 Experimental Setup

The experimental setup shown in Fig. 53 included the following equipment: a programmable virtual controller MicroLabBox DSpace (DS1202-04), two drivers with temperature controller (ITC510), a current driver (LDC240C), a temperature controller (TED200C), three function generators (AFG3021B), three laser diodes (BL976-PAG500), three EDFLs (M5), six Bragg gratings of a single-mode optic fiber, three wavelength divisor multiplexers (WD9860BA), three photodetectors (PBD481-AC), an oscilloscope (DSO-X 3102A) and a personal computer.

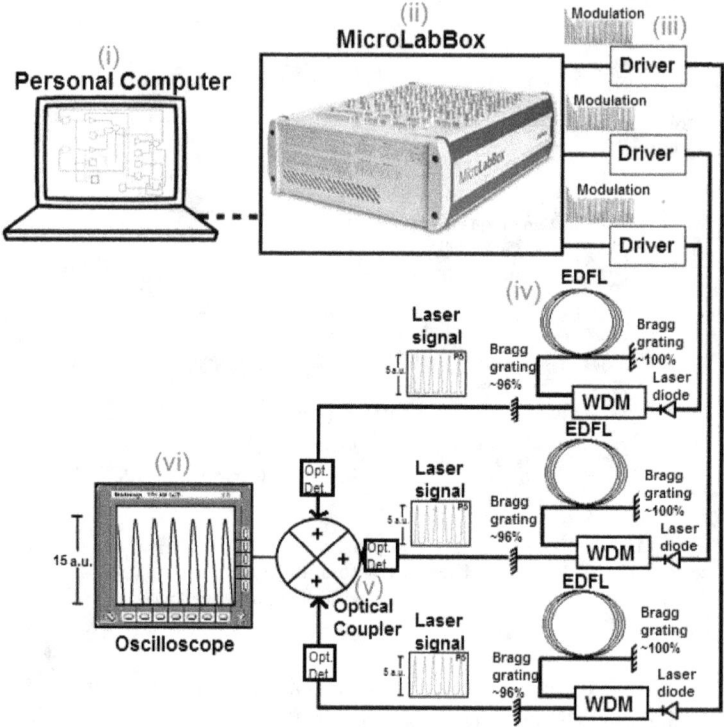

Figure 53   Experimental setup and equipment containing (i) personal computer, (ii) MicrolabBox, (iii) drivers, (vi) multistable EDFL array, (v) optical coupler and (vi) oscilloscope.

The EDFL differential Eq. (17.1) were solved using using Simulink MatLab®. We selected the values of the modulation frequency F_m and initial conditions (x_0, y_0) in the region where the system displayed multistability. In section (ii), MicroLabBox (master laser) executed the EDFL differential equations with operating parameters corresponding to the multistable regime. The communication between the computer and MicroLabBox was realized using an RJ45 cable. In section (iii), the MicroLabBox analog outputs were used to modulate each laser diode controller to pumps three EDPLs (slave lasers). This process was executed in section (iv). In section (v), three output analog signals from the EDPL array (slave lasers) were sum to obtain a giant pulse which amplitude was measured with the oscilloscope in section (vi).

17.4 Results

Each of three EDFLs is described by the same Eq. (17.1). The pump powers of the slave lasers were controlled using the coupling strength k and the output signal of the master laser ($x = x_{master}$) which modulated the intensity of each slave laser as follows

$$P_{pump} = 506(1 + m\sin(2\pi F_m t) + kx_{master}). \qquad (17.2)$$

When the coupling strength k approached a threshold value k_{th}, the slave lasers completely synchronized. Therefore, for $k > k_{th}$ the peak intensity of the sum signal obtained from the three-laser array was very high, as shown in Fig. 54.

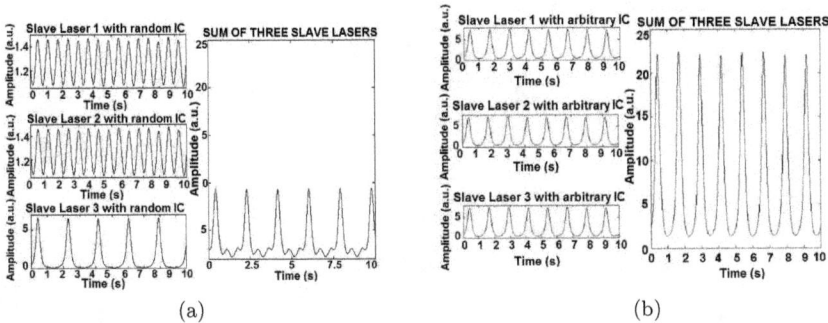

(a)   (b)

Figure 54   (Left panels) Time series of slave EDFLs 1, 2 and 3 and (right panels) sum signal for (a) $k = 0.15$ and (b) $k = 0.99$.

17.5 Conclusion

Synchronization of the array of multistable EDFLs by the master EDFL to a single periodic orbit allowed an increase of the peak intensity of the sum signal. This was achieved through a suitable choice of the coupling strength. This result was demonstrated with the bifurcation diagrams and basins of attraction of the coexistent states in the EDFL.

Acknowledgements

The authors acknowledge support from the University of Guadalajara under the project R-0138/2016, Agreement RG /019/ 2016 UdeG, Mexico. J.O.E.T. (CVU 854990) acknowledges support from the National Council of Science and Technology (CONACyT) for his Master in Science and Technology studied at CULagos in UdeG.

Symbolic representation of neuronal dynamics

Krishna Pusuluri and Andrey Shilnikov

Georgia State University, USA

18.1 Abstract

We demonstrate a GPU-based symbolic toolkit to study a whole range of dynamical behaviors occurring in neuron models. Its algorithms include periodicity detection, hashing, and Lempel-Ziv complexity to process symbolic sequences extracted from wave-form traces using voltage and time interval partitions. This aggregated partitioning scheme is well applicable to a broad spectrum of other dynamical systems across diverse disciplines. Our approach is motivated by experimental neurophysiology where voltage wave-forms are often the only observables available. This symbolic toolkit can offset and complement other computational tools for studying neuronal dynamics such as spike counting, Lyapunov exponents and parameter continuation [Barrio *et al.* (2014); Ju *et al.* (2018); Shilnikov (2012)].

18.2 Deterministic Chaos Prospector (DCP)

In our previous studies implementing such a GPU-based (multi-core graphics processing unit) symbolic toolkit, called Deterministic Chaos Prospector (DCP), we examined several Lorenz-like systems (Pusuluri *et al.*, 2017; Pusuluri and Shilnikov, 2018) to disclose a wealth of universal homoclinic and heteroclinic bifurcations of saddle equilibria, as well as to detect regions of simple and chaotic dynamics in the parameter space. Particularly, we relied on the Z_2-symmetry of such systems to generate and associate periodic

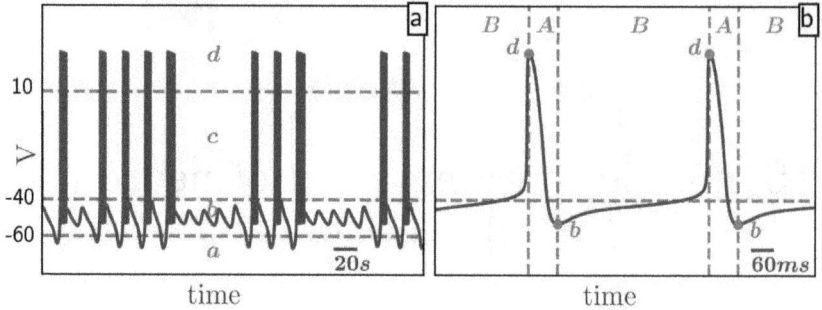

Figure 55   (a) Symbolic partitions demonstrated for mixed-mode chaotic bursting, partitioned by voltage levels at $[-60, -40, 10]$mV (red dashed lines), resulting in a set of 4 symbols ($a \leq -60 < b \leq -40 < c \leq 10 < d$). A short segment showing two spikes in a burst is magnified in (b). Based on the occurrence of events of maximal and minimal voltage values (red dots) and using voltage partitions, the symbolic representation of this segment is coded as *(dbdb)* (in red color). With time interval partitions [100]ms between events (gray dashed lines enclosing spikes) resulting in a set of 2 symbols ($A \leq 100 < B$), the symbolic representation is given by $(BABAB)$ (in gray color). Combining both voltage and interval partitions gives a detailed symbolic sequence $(BdAbBdAbB)$.

or aperiodic binary sequences, corresponding to regular or chaotic flip-flop patterns of the outgoing separatrix of the saddle at the origin. Unlike the case of the Lorenz-like attractors, chaos in slow-fast models of individual neurons is less typical. It generally occurs at transitions between bursting and tonic-spiking activity, when the corresponding orbit in the phase space changes its characteristics or stability through non-local bifurcations such as spike-adding, for example, through well-defined homoclinic bifurcations of periodic orbits and equilibria. The further development and use of the toolkit must, therefore, be motivated by a neuroscience-specific context based on the voltage wave-forms, without relying on having access to all the phase variables of a Hodgkin-Huxley model in question, such as the gating variables.

One simple method for constructing a meaningful partition to differentiate between various voltage wave-forms is to break any given one into small time-bins of an identical size, shorter than the duration of a typical spike. As these bins sample over the trace, the occurrence of a spike within a bin is marked with the symbol 1, and otherwise with 0, thus digitizing the voltage trace into a binary sequence. Alternatively, we can identify events corresponding to maximal and minimal voltage values on all spikes in the trace. Whenever a maximum is detected in the trace above some *firing* threshold, we mark it with the symbol 1, and whenever a minimum is detected below this threshold, we mark it with 0. For a typical square-wave bursting trace (without sub-threshold oscillations), this approach would be identical to spike counting. To stably identify a diverse set of neuronal dynamics including quiescent states, periodic tonic spiking, spike addition, square-wave bursting, plateau-busting, parabolic bursting, mixed-mode oscillations, quasi-periodicity and chaos, one should combine both voltage- and time-bin approaches, resulting in a minimal information loss algorithm that basically retains all relevant details, and describes the trace in the form of a multi-symbol string.

Figure 55 illustrates one of the complex bursting traces with an unpredictable number of spikes within bursts that are separated by slow amplitude sub-threshold oscillations, also chaotic. Such a trace is typically recorded in the Plant endogenous parabolic burster [Alacam and Shilnikov (2015)] at the transition between bursting activity and the hyper-polarized quiescent state. To find a symbolic description of this chaotic voltage trace, we first identify all events corresponding to maximal and minimal voltage values (red dots), as well as time intervals between them (gray dashed lines). We then use voltage and time interval partitions, V_{bins} and T_{bins}, to symbolically characterize the event and timing information. Using the voltage partition $V_{bins} = [-60, -40, 10]$mV (red dashed lines) results in four symbols ($a \leq -60 < b \leq -40 < c \leq 10 < d$), representing quiescence or burst terminations, sub-threshold oscillations, plateau burst, and spiking, respectively, found in a voltage trace. Similarly, the time interval partition $T_{bins} = [100]$ms results in a set of 2 symbols, ($A \leq 100 < B$), representing successive maximal/minimal events separated by a duration shorter or longer than 100ms, respectively. Figure 55(b) shows a short segment of two spikes in a burst within the long voltage trace. The symbolic representations of this segment using V_{bins}, T_{bins}, and a combination of both, are given by ($dbdb$) (red), ($BABAB$) (gray) or ($BdAbBdAbB$), respectively.

An overbar, like in $(\overline{abc})$, is meant to represent the periodic portion of a repetitive sequence that corresponds to regular tonic-spiking or bursting traces. For example, a tonic-spiking trace with two spikes like in Fig. 55(b) might be represented by $(\overline{db})$, $(\overline{BA})$ or $(\overline{BdAb})$, respectively, with V_{bins}, T_{bins} or combined partitions. Using just V_{bins}, spike addition to a burst starting from single spikes up to a burst with 4 spikes can be represented by $(\overline{da})$, $(\overline{dbda})$, $(\overline{dbdbda})$ and $(\overline{dbdbdbda})$, respectively. A quiescent state lacking all critical events is marked with the symbol a.

Omitting some long transient lets us examine long-term behaviors of solutions of the model in question. We normalize all shift-symmetric periodic sequences by designing a one-way hash function that produces identical hash value for all circular variations of a periodic sequence (Perlman *et al.*, 2016). In simple terms, all four circular variations of the periodic sequence $(\overline{abcd})$, $(\overline{bcda})$, $(\overline{cdab})$, or $(\overline{dabc})$ result in the same numerical hash value. For aperiodic strings representing chaotic traces, we use the LZ compression algorithm implemented in [Pusuluri and Shilnikov (2018)] for deterministic chaotic systems, to measure its complexity. As the string is scanned, new words are continuously added to the vocabulary. Eventually, the size of the LZ-vocabulary normalized by the length of the string is used as the complexity measure.

18.3   Bi-parametric Sweeps

Next, we demonstrate the benefits of this toolkit through highly detailed reconstructions of biparametric sweeps, shown in Fig. 56, for three neuronal models: (a) a mathematical three-dimensional (3D) Hindmarsh-Rose model of a square-wave burster, (b) a highly detailed 12D bull-frog hair cell model featuring quasi-periodic oscillations, and another (c) 3D leech heart interneuron model featuring various bi-stable states and the blue-sky catastrophe bifurcation on the border of tonic-spiking and bursting activity; see Refs. (Barrio and Shilnikov, 2011; Barrio *et al.*, 2014; Neiman *et al.*, 2011; Ju *et al.*, 2018; Shilnikov, 2012) and the references therein, for details of the models and parameters. The biparametric sweeps are obtained by computing long traces using identical initial conditions as the two parameters are varied across a grid of size 1000×1000. Numerical integration is performed using the fourth order Runge-Kutta method with fixed step size. The computation of these solutions is massively parallelized by running on separate GPU threads using CUDA, which results in fast computation of the sweeps such as Fig. 56(a) in about 200s. Visualizations are done in Python. A combination of V_{bins} and T_{bins} is employed to obtain symbolic sequences for events corresponding to the maximal and minimal voltage values, and/or time intervals. To study long-range dynamics of solutions of the models, a sequence of the first 2000 symbols is omitted as transient. The following sequence of 2000 symbols is then analyzed to detect the existence of periodicity (or lack thereof). If detected, the hash function generates the shift symmetric hash value of the periodic sequence, which is projected to a color map to obtain a color value. Parameter values that result in topologically

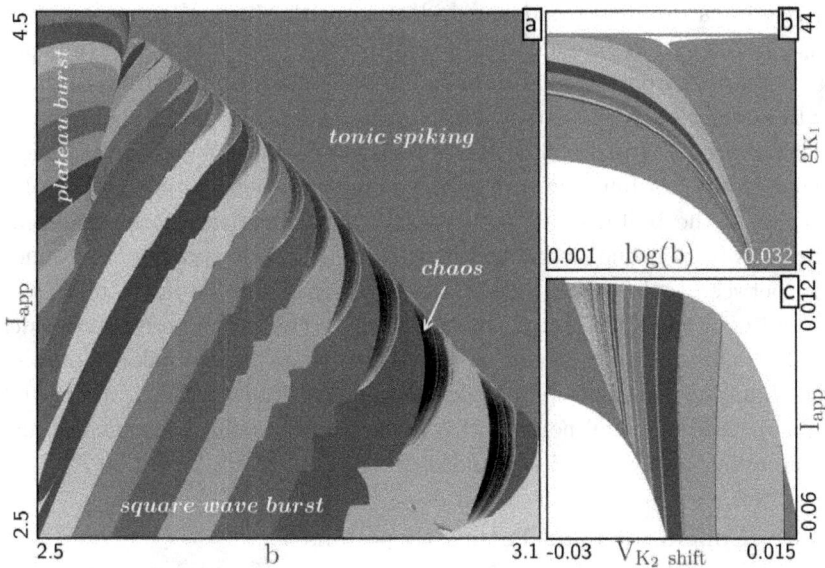

Figure 56   Bi-parametric sweeps for the Hindmarsh-Rose model (a), the bull-frog hair cell model (b), and the leech interneuron model (c) using DCP reveal universal and diverse features. The HR model (a) demonstrates plateau-bursting and square-wave bursting in addition to tonic spiking (purple) and chaos (darker shades of gray imply greater LZ complexity). The bifurcation diagram is identical with V-bins $[-1.2]$ or τ-bins [25]. Staircase-like patterns in (a) at the boundaries of spike adding transitions within the region of square-wave bursting are indicative of multi-stability. The hair cell model (b) shows tonic spiking (olive), quiescence (white), or bursting regions with spike-adding transitions between solid color stripes, same as in the bifurcation diagram of the leech interneuron model (c). Addition of noise in (c) widens the regions of chaos at the boundaries of spike addition.

identical periodic behavior in their solutions result in identical hash values, and thus, have identical color in the sweep. Aperiodic sequences are processed through the LZ-algorithm to detect their complexity. Lack of periodicity is indicative of structurally unstable, chaotic dynamics. They are represented in the bi-parametric sweeps in gray shades, with greater LZ-complexity shown in darker gray to represent greater instability.

The bi-parametric sweeps shown in Fig. 56 reveal a rich variety of dynamics in three typical neuronal models. Figure 56(a) being a sweep of the Hindmarsh-Rose model, demonstrates regions of the plateau- and square-wave bursting in addition to tonic spiking (purple) and chaos (gray shades).

The bifurcation diagram remains identical with either $V_{bins} = [-1.2]$ or $\tau_{bins} = [25]$. At the boundaries of spike-adding bifurcations, one can spot a staircase-like pattern due to bi-stability of coexisting bursting solutions with distinct spike numbers, whose emergence depends on the choice of initial conditions. Figure 56(b) shows regions of tonic-spiking (olive), quiescence (white), or bursting with spike adding transitions between solid-color stripes in the bull-frog hair cell model. The boundary of the quiescent region (white) is due to the Andronov-Hopf bifurcations, while the boundary between tonic-spiking (olive) and bursting regions is due to torus and period-doubling bifurcations (Ju *et al.*, 2018). The leech interneuron model is also known to have regions of quiescence (white), tonic-spiking (purple) and bursting partitioned by spike-adding transitions within (other solid colors). Adding small noise to this model amplifies chaos at spike-adding bifurcations and widens its boundaries [Channell *et al.* (2009)].

18.4   Conclusions and Future Directions

We showed how a novel design of multi-bin voltage and time interval partitions enhancing the previously developed DCP-toolkit expedites examinations of dynamics of simple and biologically plausible models of individual neurons, and the sweeps of their parameter spaces, to a few seconds. It also provides the flexibility of minimal loss of voltage and timing information. While this study employs manually built partitioning schemes, future development of the algorithms could apply statistical post-processing of event data to achieve scale-invariance and to enrich the high-resolution sweeps with additional temporal information concerning bursting and tonic-spiking activity. We also plan to extend these techniques for studies of the dynamics of neural networks.

Acknowledgements

This work was in part funded by NSF grant IOS-1455527, RSF grant 14-41-00044 at the Lobachevsky University of Nizhny Novgorod, and MESRF project 14.740.11.0919. We are grateful to Georgia State University's Brains and Behavior initiative for fellowship and pilot grant support and to NVIDIA Corporation for the donation of Tesla K40 GPU used in this study. We thank all the members of Shilnikov NeurDS lab for helpful discussions.

Spontaneous otoacoustic emissions from higher order signal coupling

Karlis Kanders[1,2] and Ruedi Stoop[1,2]

[1]*Institute of Neuroinformatics, University of Zürich and ETH Zürich,
Switzerland*
[2]*Institute of Building Materials, ETH Zürich, Switzerland*

19.1 Healthy Ears make Sounds

Mammalian ears emit sounds termed spontaneous otoacoustic emissions
(SOAE). These sounds, first predicted by Gold in 1948 [Gold (1948)] and
found experimentally by Kemp in 1979 [Kemp (1979)] can be detected by
placing a sensitive microphone in the ear canal and performing spectral
analysis (Fig. 57) of the recorded acoustic signal. SOAE are a prominent
signature of the active non-linear amplification process taking place in the
inner ear at the level of the outer hair cells (OHC). The inner ear has been
found to operate like a dynamical system at a Hopf bifurcation [Eguiluz
et al. (2000); Camalet and Prost (2000); Kern and Stoop (2003); Lorimer
et al. (2015)], at the critical transition from quiescence to self-sustained
oscillations. More precisely, the cochlear amplifiers are located below this
point of bifurcation [Kern and Stoop (2003)], which endows them with the
remarkable features of small-signal amplification and non-linear compression [Wiesenfeld and McNamara (1985); Derighetti *et al.* (1985); Wiesenfeld
and McNamara (1986)], resulting in remarkable 12 orders of magnitude of
dynamical range. The proximity to bifurcation instability also implies the
possibility of a crossing of the bifurcation threshold into the self-oscillation
regime. Such a crossing could occur even in the absence of any external
sound stimulation, e.g., due to the presence of (internal) sources of noise, a

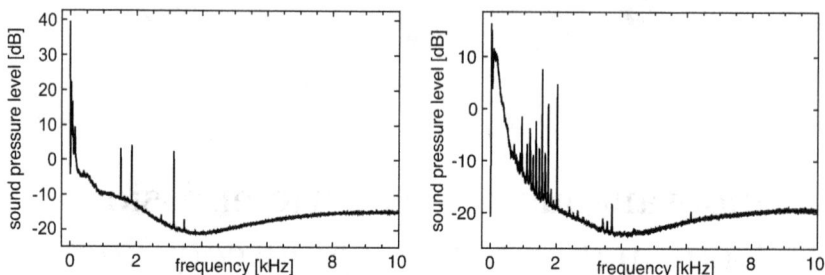

Figure 57   Examples of SOAE recorded from healthy human ears (data from [Talmadge et al. (1993)]).

mechanism that has been proposed as the likely origin of SOAE quite early [Gold (1948)] (however, cf. [Shera (2003)] for alternative hypotheses).

A mechanism that could have held responsibility for pushing the cochlear amplifiers across the bifurcation threshold remained, however, unclear. Initially, SOAE have been linked with inner ear pathology [Ruggero et al. (1983)]. While some correlation between SOAE and experimentally induced cochlear defects has been found indeed [Clark and Bohne (1984)], subsequent more extensive studies did not reveal any anatomical abnormalities associated with SOAE from normal, healthy ears [Lonsbury-Martin et al. (1988); Ohyama et al. (1991)]. Thus, SOAE are now regarded as a feature of the normally functioning ear, prevailing in approximately 70% of human population. Nonetheless, the predominant theoretical explanation of SOAE today requires the presence of some cochlear 'irregularities' and to explicitly set the cochlear amplifiers above the Hopf bifurcation [Vilfan and Duke (2008); Liu et al. (2012); Fruth et al. (2014)].

Interestingly, however, ears of females produce significantly more and stronger emissions compared to male ears [Talmadge et al. (1993); Abdala et al. (2017)] and the number and level of emissions decreases with age. This points to a different direction of the origin of the phenomenon. The novel explanation of SOAE that we put forward here builds on fewer and more natural assumptions, by identifying the effect as the potential consequence of the recently revealed *signal coupling* between the cochlear amplifiers [Gomez et al. (2016)].

19.2 Hopf Cochlea Model with Signal-coupled Feedback

If an ensemble of sub-threshold Hopf-type systems is coupled by means of their output signals, the ensemble becomes, surprisingly, more sharply tuned than the individual components. Ensembles composed of individual components that are clearly below the bifurcation point, can even undergo a collective Hopf bifurcation [Gomez *et al.* (2016)], if their coupling is strong enough. This observation led those authors to suggest that signal coupling might underlie the emergence of SOAE. To explore this hypothesis, we use a well established model of the inner ear, the Hopf cochlea [Stoop *et al.* (2007); Martignoli *et al.* (2007); Martignoli and Stoop (2010)]. This mesoscopic model of the cochlea is based on the detailed biophysics and nonlinear dynamics of the biological example [Kern (2003); Kern and Stoop (2003)] and has been found to reproduce all salient hearing characteristics (from input frequency- and level-dependent amplification profiles, to the correct combination tone generation [Gomez and Stoop (2014)] and human hearing threshold [Kanders *et al.* (2017)] and even the perception phenomena of pitch [Gomez and Stoop (2014)] and dissonance [Ignatiadis (2018)]). The model is composed of serially connected sections representing discretized parts of the cochlea (Fig. 58(a)): Each section embodies the dynamical properties of the amplification-providing OHC, well-modelled by a frequency-rescaled Hopf equation:

$$\dot{z}_j = \omega_j[(\mu_j + i)z_j - |z_j|^2 z_j - F_j(t)]; \; z_j, F_j(t) \in \mathbb{C}, \qquad (19.1)$$

where $z_j(t)$ denotes the response amplitude of the j-th section, $F_j(t)$ the forcing signal, ω_j is the section's characteristic frequency, and μ_j is the Hopf parameter. Dissipation due to fluidal viscous losses is described by tailored 6th-order Butterworth low-pass filters which complement each section directly after the Hopf system (for more details cf. references above). Conventionally, signals in the model propagate in a strictly feed-forward manner that mimics the travelling wave on the basilar membrane: The amplified and low-pass filtered response out_j is passed on to the next section as $F_{j+1}(t)$.

For modeling SOAE, we consider the potential presence of a signal-coupled feedback mechanism among the amplifiers. To implement such a feedback, we take a scaled response of the successive section, $g \cdot z_{j+1}$, and add it to the forcing signal $F_j(t)$. This coupling departs fundamentally from the more conventional 'diffusive' coupling often used in the modeling of SOAE, where interaction is proportional to the state difference between the coupled nodes. In the following, we will use Hopf cochlea parameters that have been found to correctly reproduce the human hearing threshold [Kanders *et al.* (2017)] and measure the responses for different feedback coupling strengths g. For the demonstration of the effect, we will proceed with the most naive assumption that g is the same for the whole sensor. Because we are interested in spontaneous emissions, no external input signal will be present (i.e., $F_1(t) = 0$).

19.3   SOAE Emerge from Small Signal-coupled Feedback

Signal propagation in the cochlea is essentially feed-forward; consequently, the feedback coupling strength should be relatively weak. We find that, indeed, spontaneous oscillations emerge in the cochlea already at small values of g corresponding only to a few precent of the feed-forward coupling strength (Fig. 58(b)–(f)). With increased coupling, as a trend, the number of emissions gradually increases, and the frequencies f_{SOAE} of the generated SOAE slightly drifts, and the emission spectra become increasingly complex. The cochlea response also exhibits abrupt changes (e.g., at $g \approx 0.19$ and $g \approx 0.23$), which seems to indicate a sort of collective bifurcations. In the frequency spectra, weaker and stronger emissions appear interleaved even though g is the same throughout the sensor and μ_j are very similar for adjacent sections. Some of the weaker SOAE can be attributed to cubic combination tones of other emissions. These observations are paralleled by what is found in the biological example [Burns and Jones (1984)].

The probability frequency distribution of human f_{SOAE} decays roughly exponentially. We also find this property in our model emissions, with a decay that, unfortunately, is somewhat slower. This suggests the presence of an additional process that prevents high frequency emissions. For example, there may be different conditions for signal coupling at the high-frequency end of the basilar membrane or high-frequency SOAE might be subject to some filtering process as they propagate towards the ear canal. In humans, per ear from none up to 32 emissions have been detected [Talmadge *et al.* (1993)]. In our model, only a modest variation in g is necessary to obtain such a range. A factor that could strongly influence signal coupling strength in the biological setting is OHC density. Female cochleae have typically twice as many SOAE compared to the male case, on typically shorter cochlea [Sato *et al.* (1991)] (there is, moreover, a general tendency for shorter cochleae to have a larger OHC density [Ulehlova *et al.* (1987)]). Whereas this feature is reflected in our approach, in diffusive

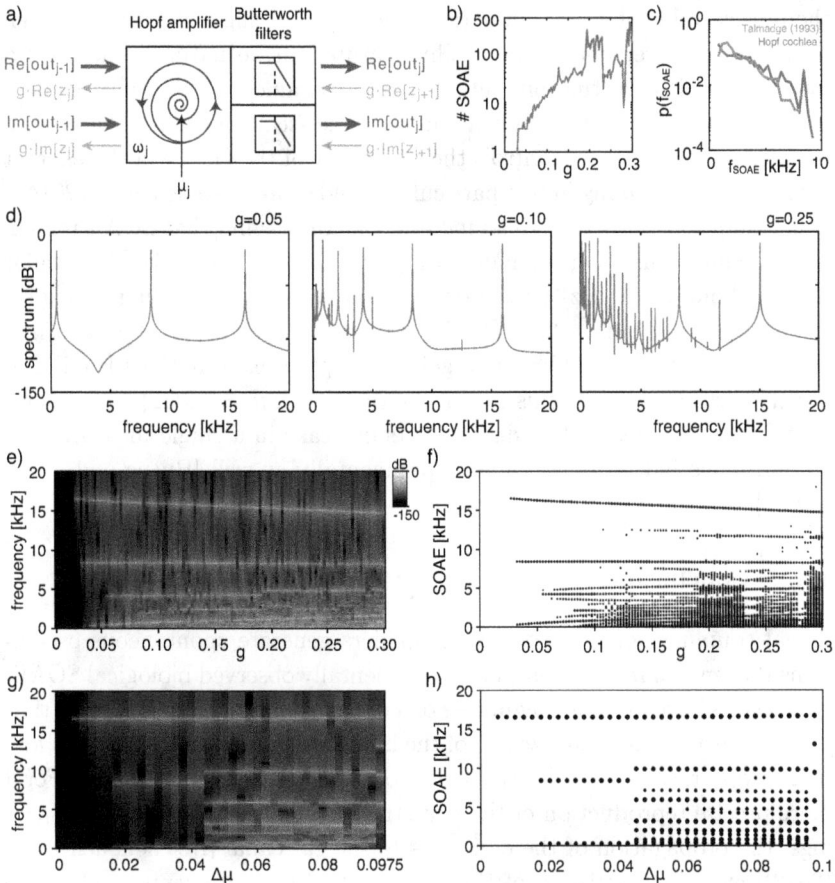

Figure 58 (a) Hopf cochlea section with signal-coupled feedback (in green) from the successive section; (b) Number of SOAE; (c) Probability of SOAE frequency for $f_{SOAE} \in (0.5, 9)$ kHz; (d) Examples of Hopf cochlea responses; (e) Power spectra and (f) SOAE for different g; (g) Power spectra and (h) SOAE when $g = 0.025$ is kept fixed, and μ_j for all sections is uniformly increased by $\Delta\mu$. Simulations were performed with 800 kHz sampling rate; SOAE were identified by summing up all sections' responses, calculating the power spectrum, and detecting peaks using 5dB threshold above baseline.

coupling frameworks, stronger coupling would, in contrast, imply less emissions [Liu *et al.* (2012)].

In our modelling framework, g is regarded as a 'structural' parameter that, over short time-scales, can be regarded as constant, an assumption that can be justified from the observation that in an individual ear the same emissions can be found for two decades [Burns (2009)]. However, the precise value of f_{SOAE} may (and does) exhibit slight fluctuations even during a measurement session. This feature can be naturally accounted for by including fluctuations into the Hopf parameters μ_j. Such changes are expected to occur on much shorter times-scales because the central nervous system actively controls the sensitivity of the OHC via the efferent connections, for tuning in to a particular sound source [Gomez *et al.* (2014)]. When all μ_j are subjected to an increase by $\Delta\mu$, we find that SOAE remain robust while exhibiting a small drift (Fig. 58(g)–(h)) that is of a similar range (about 10–50 Hz) as the one measured in corresponding biological experiments [Ohyama *et al.* (1991); Norton *et al.* (1989)]. We also find a rare occurrence of SOAE undergoing abrupt transformations leading to largely distinct sets of emission peaks (around $\Delta\mu \approx 0.042$). Emissions that fluctuate between two distinct sets of peaks in a single measurement session, have indeed been found experimentally as well [Burns and Jones (1984)].

19.4   Conclusions

Signal coupling between cochlear amplifiers generate spontaneous oscillations that match in many aspects experimentally observed biological SOAE. Our results have provided a glimpse of a complex and rich phenomenon that invites a further characterization of the influences from topology, delays and node properties on the collective behavior of signal-coupled networks. For a more precise reproduction of the statistics of the emissions and their spacings, the propagation of the emissions to the ear canal [Reichenbach *et al.* (2012)] and the sensitivity of the experimental measurements might have to be included. The insights obtained in this study promise already at the present stage, to be relevant not only to the hearing system, but other biological networks as well.

Acknowledgements

The authors thank Tom Lorimer for insightful and useful discussions in the initial stages of this research. The authors express their gratitude to C.L. Talmadge for kindly providing the experimental data reported in [Talmadge *et al.* (1993)] and useful comments regarding the said dataset. The research was supported by an internal grant of ETH Zürich, ETH-37 152.

A physics-driven human brain alternative

Karlis Kanders[1,2] and Ruedi Stoop[1,2]

[1]*Institute of Neuroinformatics, University of Zürich and ETH Zürich, Switzerland*
[2]*Institute of Building Materials, ETH Zürich, Switzerland*

20.1 The Human Brain Alternative

Each century has its own major scientific focus; the present century is likely to be called the 'brain age', referring to the humanity's joint effort to understand not — as previously — its objective physical environment, but how it perceives the latter. To achieve this, the reconstruction of the physiological brain in terms of hard- or software is the main strategy. This attempt requires huge investments of scientific efforts, animal lives, and laboratory and computer equipment. In our contribution, we point out that understanding the physics principles and their consequences relevant for the functioning brain, may offer from a complementary side a more efficient strategy to achieve this goal.

The alternative is, more specifically, to consider the human mind as an onion made up by shells. Across these shells, we follow the signal transformations and transductions, successively from the sensors towards the innermost shells of the human outer world perception. In this way, we peel the onion from the outside shells to innermost stages, progressing towards what we actually mean by the term 'brain'. Such a focus on technical, nonlinear sensory properties promises, moreover, more direct technical applications compared to the more biologically driven mainstream neuroscience approach.

Hampered by an only rudimental understanding of the human sensors, many properties of our perception of the world have been dogmatically assigned to the 'brain'. For providing an example of this alternative approach, we will be focusing on the example of human hearing as arguably most complex perception of the physical processes in our environment (which may only be challenged by the mammalian perception of smells, which — discounting involved chemistry of the receptors — we expect, however, to follow essentially the example of hearing).

With a focus on varying species-specific frequency intervals, mammalian hearing is able to access a huge dynamic range of sound (between 120–130 dB). This is due to the ability of the cochlea's outer hair cells to generate nonlinear amplification of the incoming sound signal, leading to strong amplification of weaker sounds and weaker amplification of stronger sounds [Ruggero *et al.* (1997)]. Outer hair cells follow in physical and in frequency space (their connection is expressed by the tonotopic map) a largely scaled building plan [Lorimer *et al.* (2015)]. Traditionally, the mammalian hearing sensor, the cochlea, has been described at several levels. The finest one is the level of the outer hair cells, where often the focus is put on explaining the intriguing interaction between hair bundle mechanics and electromotility of the hair cell bodies. On a more mesoscopic level, the cochlea's building plan can be captured in terms of so-called Hopf amplifiers (e.g., [Lorimer *et al.* (2015)]). The sensor can then be composed as a sequence of mesoscopic sections representing discretization parts of the cochlea, yielding the so-called Hopf cochlea (e.g., [Martignoli *et al.* (2007)]), where the sections and their succession are based on the detailed biophysics and nonlinear dynamics at work in the cochlea [Kern and Stoop (2003); Kern (2003)]. Fundamental for this model is that the sections share the dynamical properties of the microscopic amplification-providing outer hair cells [Kern (2003); Gomez *et al.* (2016)], which are captured in a stimulated *Hopf process*

$$\dot{z} = \omega_{ch}((\mu + i)z - |z|^2 z - F(t)); \; z, \, F(t) \in \mathbb{C},$$

where $z(t)$ denotes the response amplitude, $F(t)$ a stimulation signal, ω_{ch} is the characteristic frequency of the Hopf system, and μ is the *Hopf parameter* [Eguiluz *et al.* (2000); Camalet and Prost (2000); Kern and Stoop (2003); Kern (2003); Stoop and Gomez (2016)]. At values $\mu < 0$, the system is below bifurcation to self-oscillation, but responds towards stimulation signals $F(t)$ as a small-signal amplifier [Wiesenfeld and McNamara (1985); Derighetti *et al.* (1985); Wiesenfeld and McNamara (1986)]. The main characteristics of the isolated node dynamics are collected in

Ref. [Stoop and Gomez (2016)]. Essentially, the local value of the parameter μ embraces the main features of the basilar membrane plus the amplification provided locally by the outer hair cells, whereas the dissipation by fluidal viscous losses can be described by tailored 6th-order Butterworth low-pass filters [Martignoli *et al.* (2007); Martignoli and Stoop (2010)]. The biophysical properties of the cochlea suggest selecting the characteristic frequencies of the nodes according to a geometric sequence. When embedded into a compound cochlea, due to the sections' interaction with neighboring ones, the response profiles broaden and reproduce, with properly chosen μ-parameters, the measurements of the biological cochlea [Ruggero (1992)] extremely well [Martignoli *et al.* (2007)].

We generally use a software implementation of an earlier hardware realization of 29 sections or nodes, taking care of 7 octaves ($14.08 - 0.11$ kHz), or a 31-section model covering an interval of ($19.912 - 0.11$) kHz. Our partition is optimal in the sense that finer partitions yield for the human amplification range, identical results, but coarser partitions lead to distortions in the frequency dependence, if sufficiently strong amplification is required. For *'flat tuning'*, $\mu \equiv$ const, all nodes would have identical Hopf parameters (conventionally $\mu \equiv -0.25$) [Stoop and Gomez (2016)], but to optimally match the human hearing system, we implement a soft continued gradual amplification decay towards lower frequencies. On this basis, we may tune their values actively, to mimic the effect of efferent cochlear connections in the context of *learning* [Gomez *et al.* (2014)]. Conditioning the network towards desired sounds, by tuning nodes that are unhelpful towards weaker amplification, we achieve a remarkable 'listener' effect, i.e. an ability of the system to extract and follow desired sound sources in mixtures of sounds.

Note that our cochlea model focuses on the active amplification by the inner ear. In addition to active amplification, other processes influence hearing, in particular at high SPL (e.g., bone conductance) or at low frequencies (vibrotactile excitation). These effects are not the subject of our model.

20.2 Reproduction of Psychoacoustic Phenomena

Many puzzling human psychoacoustic phenomena were originally assigned to the brain. In the recent past, they, however, clearly emerged as the properties of the mammalian hearing sensor, the cochlea. As was mentioned above, across the whole extension of the cochlea, a down-regulation of the sensitivities μ becomes generally non-negligible [Kanders *et al.* (2017)].

However, the majority of the biological measurements that theoretical models can be compared with, refer to local properties only, involving nearest neighboring sections at most. Across this distance, the detuning is without a noticeable effect. Phenomena that are based on local properties embrace the local amplification profiles, where the basilar membrane displacement relative to the sound pressure level at the eardrum measured for several mammals [Ruggero (1992)] could be reproduced by the Hopf cochlea [Stoop *et al.* (2007)], as were the corresponding single tone gain profiles ([Robles and Ruggero (2001); Ruggero *et al.* (1997); Cooper and Rhode (1997)] vs. [Martignoli and Stoop (2010)]). These measurements are fundamental, as they express the basic nonlinearity properties of the auditory amplifier that are responsible essentially for all of the puzzling phenomena of hearing. The compression of strong sounds exhibiting an exponent of compression of $1/3$ is taken as a strong indicator for the usage of the standard Hopf system for cochlear modeling. Mutual compression of neighboring tones is another nonlinear feature that is characteristic of human hearing. The effect can be considered as a prototype of computation done by the mammalian hearing sensor [Stoop and Kern (2004)]. In the past, some primitives of these phenomena were obtained from the amplifiers alone, but a stronger conclusion also requires to show that the amplifier properties can shine through the whole cochlea, which involves, in addition of the amplifiers, also the properties of the cochlear fluid. The nonlinearities in the amplification process also introduce, by means of amplifier interaction, additional tones called combination tones, that are also reproduced by the model (cf. [Robles *et al.* (1997)] vs. [Martignoli and Stoop (2010)]). Such tones — absent in the physical stimulation outside of the ear — can in fact be heard by a careful listener. Combination tones (cf. [Robles *et al.* (1997)] vs. [Martignoli and Stoop (2010)]), and their decay laws (cf. [Robles *et al.* (1997)] vs. [Gomez and Stoop (2014)]), are of great importance for the human perception of pitch.

Global properties of the cochlea that have been verified by our model are the phase behavior along the cochlea, the loudness dependence of pitch [Martignoli and Stoop (2010)], and the so-called pitch shift effect (cf. [Gomez and Stoop (2014)]). The latter effect describes the human perception of the pitch generated by two tones if one of them is shifted in frequency with respect to the other's frequency [Gomez and Stoop (2014); Smoorenburg (1970); Rhode (1995)]: The psychoacoustic perceived pitches contradict the phenomenological expectation based on the nature of the stimulating sound, i.e. the famous de Boer's formula [de Boer (1976)]. The

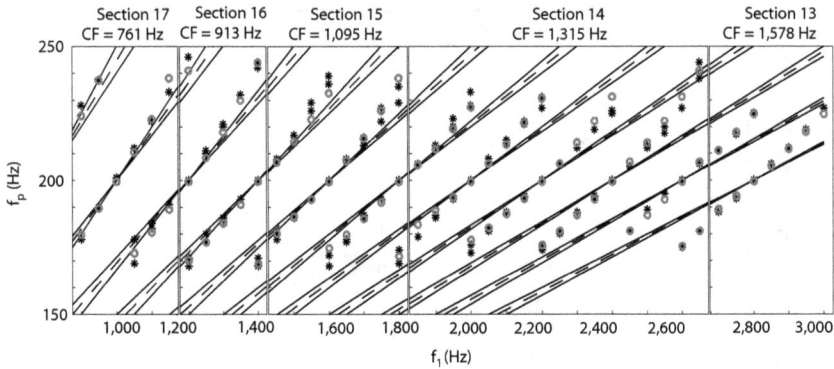

Figure 59   Pitch-shift experiment (from [Gomez and Stoop (2014)]): Two-frequency stimulation $f_2 = f_1 + 200$ Hz. Black stars: psychoacoustic data [Smoorenburg (1970)] (partial sound levels 40 dB sound pressure level, two subjects). Red circles: Hopf cochlea (sections as indicated, tones at -74 dB each). Black lines: false predictions by de Boer's formula [de Boer (1976)] for $k' = k$, $k' = k + 1/2$ (dashed) and $k' = k + 1$, respectively.

in-depth analysis of our model, supported by psychoacoustic findings, leads finally to the correct prediction of the pitch-shift and provides an explanation of the puzzle (Fig. 59). Our biologically motivated tuning of the Hopf amplifier sensitivity along the cochlea [Kern (2003)] also explains the particularities of the human hearing threshold in a more convincing manner [Ruggero and Temchin (2002); Kanders *et al.* (2017)] than the earlier, but still common, outer and middle ear-based explanations (Fig. 60).

An important verification was that of the medial efferent inhibition: The effect of a tuning of the cochlea by efferent medial olivocochlear stimulation. In [Gomez *et al.* (2014)], we showed that this efferent mechanism provides a powerful mechanism of tuning the cochlea towards desired sounds — a possibility that present hearing aids fail to foresee.

As the final stage of our reproduction of psychoacoustic phenomena, we recently revealed a cochlea-based model for 'consonant' vs. 'dissonant' sound perception. For this, we considered the cochlea as a network of a subset of activated nodes (nodes that are activated above hearing threshold [Stoop and Gomez (2016)]) and estimated the complexity of this network [Ignatiadis (2018)]. This complexity provides a quite accurate measure for the perception of dissonance, where the interpretation is that the more complex the network is, the lesser we like the sound (Fig. 3).

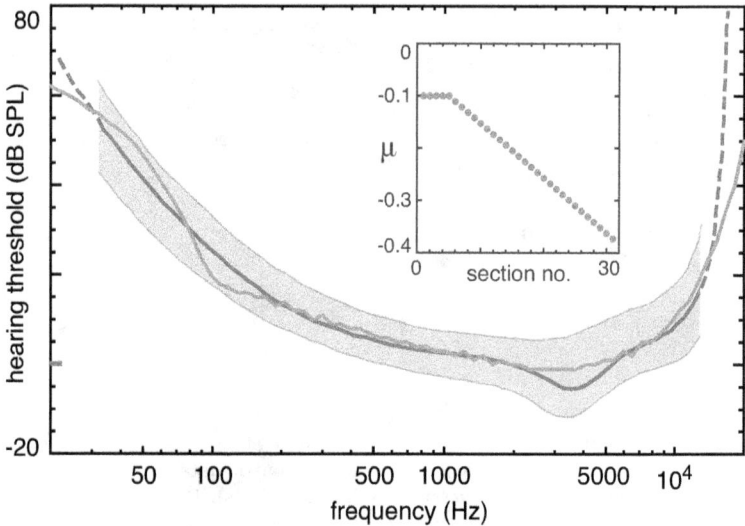

Figure 60   (Following Ref. [Kanders *et al.* (2017)]) Gray curves: Data from Zwicker's publication [Zwicker and Heinz (1995)], dashed part indicating extrapolations. Gray shading: Observed human variability [Fastl and Zwicker (2007)]. Red curve: Cochlea response with biological tuning (31 sections [Gomez and Stoop (2014)]). Across sensible tuning variations, the overall behavior is remarkably stable, and discretization effect by the last amplifier is clearly visible. Inset shows down-regulated sensitivity towards low frequency necessary for the precise correspondence, as a consequence of biophysical parameter variations captured in the Hopf parameter μ. The high sensitivity around 4000 Hz is the effect of the outer and middle ear funnel channel, the deviation around 100 Hz is an effect of the discretization of our model (no sections below) and can be mended by a finer discretization.

20.3   Conclusion

In conclusion, we have shown that very delicate facets of the human perception of the outside world is — we would say: much more than in the visual system — a direct product of the hearing sensor properties, the cochlea. The phenomena of hearing acquired at this — pre-cortical — stage are, strangely enough, as faithfully as possible transmitted to the cortex. This includes in particular the nonlinear effect, which to faithfully transmit towards the cortex the biological system takes great care, including the application of a trick of stochastic resonance [Martignoli *et al.* (2013)]. This is a surprising difference of engineering; the classical technical engineering approach would primarily have used filters to get rid of, e.g., the combination tones that, as a product of the nonlinearities, leads to a puzzling

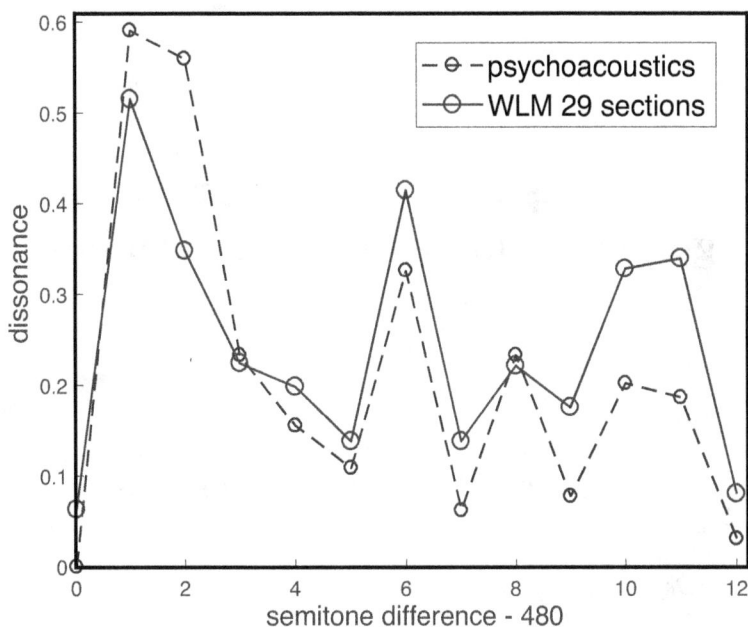

Figure 61   Network-based dissonance measure WLM, compared to the psychoacoustic perception (complex-tone inputs, 480 Hz octave base tone, cochlea of 29 sections).

complexification of the auditory information. In this sense, our approach leads us not just closer to the human perceptions previously ascribed to the cortex, but also offers an engineering approach that is alternative to the inherent narrow-ranged linearization around a previously chosen working point that is doomed to not be able to deal with the full range of sounds of the world around us, in contrast to what is our own biological system able to perform.

Acknowledgements

The research was supported by an internal grant of ETH Zürich, ETH-37 152. Before everyone else, we are indebted to Drs. A. Kern, J.-J. v.d. Vyver, S. Martignoli and F. Gomez, whose studies have paved the way for this contribution.

Consensus clustering approach for genome-wide association studies

Javier Rasero[1], Teresa Creanza[2], Nicola Ancona[2], Jesus M. Cortes[1−3],
Daniele Marinazzo[4] and Sebastiano Stramaglia[5]

[1] *Biocruces Health Research Institute, Bilbao, Spain*
[2] *ISSIA-CNR, Bari, Italy*
[3] *Ikerbasque, Bilbao, Spain*
[4] *Department of Data Analysis, Ghent University, Belgium*
[5] *University of Bari and INFN, Sezione di Bari, Italy*

21.1 Introduction

In the supervised analysis of gene expression profiling, aiming at identifying differentially expressed genes (pathways) subjects are usually grouped in correspondence of high-level clinical categories (e.g., patients and controls), and typical approaches aim at deducing a decision function from the labeled training data.

However, the population of healthy subjects (as well as those of patients) is usually highly heterogeneous: stratification of groups may be a useful preprocessing stage, so that the subsequent supervised analysis might exploit the knowledge of the structure of data. Stratification of groups may be done using phenotypic variables of subjects, if available. More interestingly, stratification may rely on the gene expression data itself, by application of clustering algorithms that find natural groupings in the data.

An effective supervised approach, named Multivariate Distance Matrix Regression (MDMR), has been proposed in [Zapala and Schork (2006)] for the analysis of gene expression patterns; it tests the relationship between variation in a distance matrix and predictor information collected on the samples whose gene expression levels have been used to construct the matrix. Whilst MDMR has found wide application, its statistical power may be certainly affected by the heterogeneity of classes.

Recently an unsupervised method [Rasero *et al.* (2017)] has been developed where different features are not combined in a single vector to feed the clustering algorithm; rather, the information coming from the various features are combined by constructing a *consensus* network. Consensus clustering is commonly used to generate stable results out of a set of partitions delivered by different clustering algorithms (and/or parameters) applied to the same data; in [Rasero *et al.* (2017)], instead, the consensus strategy was used to combine the information about the data structure arising from different features so as to summarize them in a single consensus matrix, which not only provides a partition of subjects in communities, but also a geometrical representation of the set of subjects.

The purpose of this work is to review a recently proposed approach [Rasero *et al.* (2018)] where the consensus clustering approach is seen as a preprocessing stage of MDMR in exploratory analysis, so as to cope with the heterogeneity of subjects. We will show that extracting the natural classes present in data and subsequently performing the supervised analysis between the subgroups found by consensus clustering, allows to identify variables whose pattern is altered in group comparisons, which are not identified when the groups are used as a whole. We present application of the proposed method to microarray data where the goal is to identify the pathways which are significantly altered due to the pathology.

21.2   Materials and Methods

The dataset consists of microarray data from multiple sclerosis patients and healthy controls. We used a dataset publicly available through the Gene Expression Omnibus (GEO) repository [Barrett *et al.* (2005)] with the accession number GSE41848 [Nickles *et al.* (2013)]. In detail, we analyze gene expression profiles of whole-blood RNA from a cohort of 37 multiple sclerosis (MS) patients and 79 healthy controls. The genome association analysis of this data set has been already performed using an enrichment pathway analysis in order to reveal among a set of predefined gene sets those sets that are significantly enriched for genes differentially expressed (DE) relative to two different phenotypic states, using the random set (RS) enrichment method described in [D'Addabbo *et al.* (2011); Abatangelo *et al.* (2009)]. We collected the top 50 significantly deregulated gene sets (according to the random set method) with p-value ≤ 0.02 and 50 gene sets that did not result significantly deregulated because the relative p-values were in the range [0.49, 0.506]: therefore 100 pathways were considered in the present study. Methodologically, for each pathway, we define a distance matrix $D \equiv (d_{\alpha}\beta) = \sqrt{2(1 - r_{\alpha\beta})}$, where $r_{\alpha\beta}$ is the Pearson correlation among α and β subject's gene expression for that pathway. Subsequently, the consensus clustering proposed in [Rasero *et al.* (2017)] is applied to this set of distance matrices as follows:

(1) Cluster each of distance matrix using k-medoids for range involving different k. In this particular case, we have considered the interval [2,20].

(2) Build the consensus network $C \equiv (c_{\alpha\beta})$ from the corresponding partitions, such that each entry indicates the number of partitions in which subjects α and β are assigned to the same group averaged over the the resolution parameter k and the pathways.

(3) Evaluation of the modularity matrix as $\mathbf{B} = \mathbf{C} - \mathbf{P}$, where P is the co-assignment matrix, which aims at providing the null model obtained through repeatedly label shuffling. The number of shufflings has been taken equal to 100.

(4) Group of subjects extraction by means of a modularity optimization algorithm applied to this modularity matrix.

Group-level statistical differences have been calculated by means of MDMR, which allows us to identify altered pathway's gene expression with respect to group label.

21.3 Results

The application of MDMR to the full group of healthy subjects and MS patients provides 24 significant pathways. Remarkably, 10 out of these significant results are not observed by random set analysis in [D'Addabbo *et al.* (2011); Abatangelo *et al.* (2009)].

Since the healthy population is in general highly heterogeneous, giving rise to great variability in group-comparison studies, we applied the consensus clustering to the healthy group, for which the modularity matrix in Fig. 62 is obtained, exhibiting four clusters of 13, 21,11 and 32 subjects respectively. After this group stratification, the intra-group distance distribution for each cluster decreases significantly w.r.t whole group (panel B in Fig. 62).

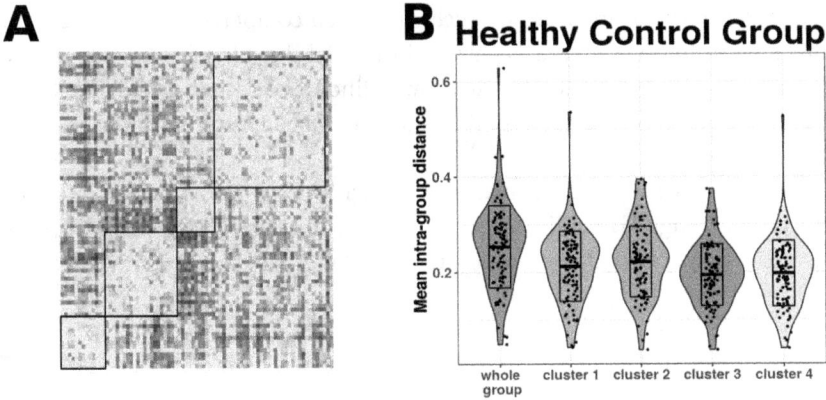

Figure 62 (Panel A) Modularity matrix. (Panel B) The intra-subjects distance distribution for healthy group and clusters found in it by our method. The distribution elements correspond to the off-diagonal entries in the collection of distance matrix given by each gene pathway.

At the cluster level when compared with the MS group, the first cluster provided 49 significative pathways, the second 3, the third one 26 and the last one 29. All these results have been corrected for multiple comparisons, controlling for False discovery rate at a 5% significance level.

Furthermore, application of consensus clustering to get more compact sub-groups allowed us to elucidate pathways invisible neither in Random set analysis nor in whole group comparisons. Such pathways are associated with mitochondrial atp synthesis and go-actin polymerization, thus suggesting an impaired function of both the ATP synthesis and the proton transport activity of the enzyme. Mitochondrial damage in MS was found to play an important role in the progression of the disease. We remark that the mitochondrial pathway was not recognised as significant by random set analysis and neither by MDMR using the full group of HC, *i.e.*, the pre-processing by consensus clustering has been necessary to highlight the role of mithocondrial damage in this data-set.

21.4   Conclusions

In this work we have made use of the consensus clustering approach to extract the natural classes present in the healthy population. As found in [Rasero *et al.* (2018)], we have demonstrated that this leads to a variability reduction in the healthy group, such that when compared to the MS group, this aids in finding altered pathways in the ATP synthesis and the proton transport activity of the enzyme. Such findings are not found using the whole group, evincing the usefulness of consensus clustering preprocessing in microarray data.

We acknowledge the University of Bari "Aldo Moro" for funding this research through Fondi di Ateneo per la Ricerca: Quota Progetti di Ricerca su base competitiva, f.di ex 60% - residuo 2015.

Synchronization of spike-trains in a coupled system of digital spiking neurons

Hiroaki Uchida and Toshimichi Saito

Hosei University, Japan

22.1 Introduction

Spiking neurons are analog dynamical systems and can generate various periodic/chaotic spike-trains. Analysis of spike-trains is important to consider interesting nonlinear phenomena and information processing function in the neural systems [Izhikevich (2006)]. Also, the spike-trains play important roles in engineering applications including image processing [Campbell *et al.* (1999)], spike-based communication [Rulkov *et al.* (2001)], and central pattern generators [Lozano *et al.* (2016)]. Analysis and synthesis of spiking neurons are important from both fundamental and application viewpoints.

We introduce a digital spiking neuron (DSN [Torikai *et al.* (2008)] [Saito *et al.* (2017)]) and a ladder-type network of digital spiking neurons (LDSN). The DSNs/LDSNs are digital dynamical systems and can be regarded as a digital version of the analog spiking neurons/networks. The DSNs/LDSNs are suitable for precise numerical analysis and FPGA based hardware implementation. Repeating integrate-and-fire behavior between periodic base signal and constant threshold, the DSN can output a spike-train. Since the DSNs are defined on a discrete phase space of a finite number of points, the DSN cannot generate chaos but various periodic spike-trains (PSTs). Applying spike-based cross-coupling, the body of LDSN is constructed. Depending on parameters, the LDSN can exhibit multi-phase synchronization of various PSTs [Uchida and Saito (2017)]. Using shift registers and switches, the LDSN can be realized. Applying a time dependent selection switching, the LDSN can output a variety of PSTs consisting of desired inter-spike-intervals. Such PSTs are applicable to spike-based time series approximation/prediction and the LDSN can be developed into a digital version of reservoir computing systems [Appeltant *et al.* (2011)].

22.2 Digital Spiking Neuron and Periodic Spike-Train

First, we introduce the DSN that is a building block of the LDSN. Let $x(\tau)$ denote a discrete state variable at discrete time τ. Repeating integrate-and-fire dynamics between a periodic base signal $b(\tau)$ with period N_p and a constant threshold N_x, the DSN outputs various periodic spike-train $y(\tau)$. The dynamics of DSN is described by the following two types of firing:

$$\begin{aligned}
&\text{Integrating:} \quad x(\tau+1) = x(\tau) + 1, \ y(\tau) = 0 \ \text{ if } x(\tau) < N_x \\
&\text{Self-firing:} \quad x(\tau+1) = b(\tau), \qquad y(\tau) = 1 \ \text{ if } x(\tau) = N_x
\end{aligned} \tag{22.1}$$

where $x(\tau)$ is a discrete state variable, $y(\tau)$ is an output spike-train, and $b(\tau)$ is q base signal with period N_p. For simplicity, we assume the following:

$$\begin{aligned}
&x(\tau) \in \{1, 2, \cdots, N_x\}, \ N_x \le 2N_p - 1, \\
&\tau - 2N_p + 1 \le b(\tau) - N_x \le \tau - N_p \ \text{for } \tau \in \{0, \cdots, N_p - 1\}.
\end{aligned} \tag{22.2}$$

In this case, one spike appears per unit interval of N_p points.

$$y(\tau) = \begin{cases} 1 & \text{for } \tau = \tau_n \\ 0 & \text{for } \tau \ne \tau_n \end{cases} \quad \tau_n \in [(n-1)N_p + 1, nN_p]. \tag{22.3}$$

where τ_n denotes the n-th spike-position. Let $\theta_n = (\tau_n \bmod N_p)$ denote the n-th spike-phase. A spike-position is given by $\tau_n = \theta_n + N_p(n-1)$ and a spike-train $y(\tau)$ is governed by the digital spike map (Dmap)

$$\theta_{n+1} = F(\theta_n) \equiv \theta_n - b(\theta_n - 1) + (N_x - N_p), \ \theta_n \in \{1, \cdots, N_p\} \equiv L_{N_p} \tag{22.4}$$

Any periodic spike-trains described by the Dmap can be generated by a DSN [Saito *et al.* (2017)].

22.3 Ladder-Type Digital Spiking Neural Network

Figure 63(a) shows the body of the LDSN: M pieces of DSNs are coupled in ladder topology with a common base signal. The dynamics is described by

Integrating: $x_i(\tau + 1) = x_i(\tau) + 1$, $y_i(\tau) = 0$ if $x_i(\tau) < N_x$

Self-firing: $x_i(\tau + 1) = b(\tau)$, $y_i(\tau) = 1$ if $x_i(\tau) = N_x$

Cross-firing: $x_{j+1}(\tau + 1) = N_x - N_p + 1$, $z_j(\tau) = 1$ if $x_j(\tau) = N_x$

$$\text{and} x_{j+1}(\tau) \le N_x - N_p \tag{22.5}$$

where $i \in \{1, \cdots, M\}$, $j \in \{1, \cdots, M - 1\}$ and the cross-firing is prior to the integrating. The integrating and self-firing are the same as the single DSN in Eq. (22.1). The cross-firing connects DSNs in ladder topology: the 1st DSN can apply the cross firing to the 2nd DSN (see Fig. 63(a)) and the $(j-1)$-th DSN can apply the cross-firing to the j-th DSN where $j = 2 \sim M$. The self-firing is characterized by the spike-train $y_i(\tau)$. The cross-firing is characterized by the connection signal $z_i(\tau)$.

As a typical phenomenon of LDSN, we define the M-phase synchronization (M-SYN) with period MN_p.

$$x_i(\tau) = x_i(\tau + MN_p), y_i(\tau) = y_i(\tau + MN_p), i \in \{1, \cdots, M\}$$
$$x_j(\tau) = x_{j+1}(\tau + N_p), y_j(\tau) = y_{j+1}(\tau + N_p), j \in \{1, \cdots, M - 1\} \tag{22.6}$$
$$z_i(\tau) = 1 \text{ for some } \tau \in \{1, \cdots, MN_p\}$$

Existence of non-zero connection signal $z_i(\tau)$ means existence of the M-SYN. If $z_i(\tau) = 0$ for all τ, all the DSNs are isolated without coupling. Figure 63(a) illustrates an M-SYN of PSTs with period MN_p for $M = 5$.

Applying time-dependent selection switches S_i, the output of the LDSN is given as illustrated in Fig. 64. For simplicity, we consider the case where each DSN outputs a periodic spike-train with period MN_p and the LDSN consists of M pieces of DSNs. We divide the time interval $(0, MN_p]$ into M slots I_1 to I_M:

$$I_1 = [0, N_p), I_2 = [N_p, 2N_p), \cdots, I_M = [(M - 1)N_p, MN_p).$$

The i-th time-dependent selection switch S_i, $i = 1 \sim M$, selects either DSN in each time slot and gives the output as the following:

$$\text{Output: } y(\tau) = w_{ij} y_i(\tau), w_{ij} = \begin{cases} 1 & \text{if } y_i \text{ is selected for } \tau \text{ mod } MN_p \in I_j \\ 0 & \text{otherwise} \end{cases}$$

where $i = 1 \sim M$. For each time slot I_j, only one DSN is selected. The output is characterized by selection matrix $W = (w_{ij})$. Figure 64 shows an

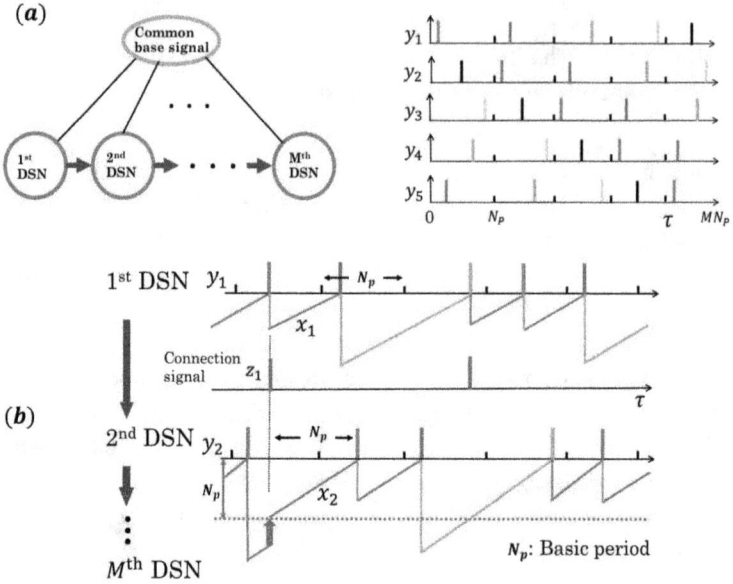

Figure 63   LDSN. (a) M-SYN of PSTs with period MN_p (M=5). (b) Cross-firing.

output for selection matrix

$$W = \begin{pmatrix} 1 & 0 & 0 & 0 & 0 \\ 0 & 1 & 0 & 0 & 0 \\ 0 & 0 & 0 & 0 & 1 \\ 0 & 0 & 1 & 1 & 0 \\ 0 & 0 & 0 & 0 & 0 \end{pmatrix} \qquad (22.7)$$

Using the selection matrix, the LDSN can output various PSTs consisting of any combination of M spike-phases. The LDSN can generates either of M^M PSTs. For example, the LDSN in Fig. 64 generates a PST of 3 spike-phases $\{\theta_a, \theta_a, \theta_e, \theta_a, \theta_c\}$. Such a PST is impossible in the DSN governed by the Dmap $\theta_{n+1} = F(\theta_n)$. Note that the PST can be stabilized by adjusting the base signal and the local stability of M-phase synchronization can be analyzed theoretically. Using the Verilog, a FPGA based test circuit can be designed for observation of typical phenomena [Uchida and Saito (2017)].

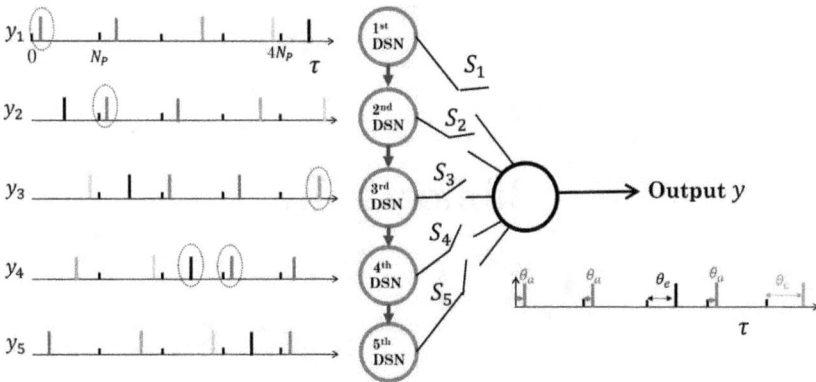

Figure 64   The output of the LDSN by time-dependent selection switches.

22.4   Conclusions

The LDSN is introduced and multi-phase synchronization phenomena of various PSTs are considered. Applying the time dependent selection switches, the LDSN can output various PSTs consisting of any combination of desired inter-spike intervals. Future problems include detailed stability analysis of multi-phase synchronization of PSTs and application to spike-based time series approximation/prediction.

Bibliography

802.15.6-2012, I. S. (2012). *IEEE Standard for Local and metropolitan area networks-Part 15.6: Wireless Body Area Networks* (IEEE Press).

Abatangelo, L., Maglietta, R., Distaso, A., D'Addabbo, A., Creanza, T. M., Mukherjee, S., and Ancona, N. (2009). Comparative study of gene set enrichment methods, *BMC Bioinformatics* **10**, 1, p. 275, doi:10.1186/ 1471-2105-10-275.

Abdala, C., Luo, P., and Shera, C. (2017). Characterizing spontaneous otoacoustic emissions across the human lifespan, *J. Acoust. Soc. Am* **141**, pp. 1874– 1886.

Addo, P. M., Billio, M., and Guegan, D. (2013). Nonlinear dynamics and recurrence plots for detecting financial crisis, *The North American Journal of Economics and Finance* **26**, pp. 416–435.

Adoul, P. (1974). Error intervals and cluster densities in channel modeling, *IEEE Trans. Inform. Theory* **20**, pp. 125–129.

Aihara, K., Itob, K., Nakagawac, J., and Takeuchiad, T. (2014). Optimal control laws for traffic flow, *Applied Mathematics Letters* **26**, pp. 617–623.

Alacam, D. and Shilnikov, A. (2015). Making a swim central pattern generator out of latent parabolic bursters, *International Journal of Bifurcation and Chaos* **25**.

Andreev, A., Makarov, V., Runnova, A., Pisarchik, A., and Hramov, A. (2018). Coherence resonance in stimulated neuronal network, *Chaos Sol. and Fract.* **106**, pp. 80–85.

Andreev, A., Pitsik, E., Makarov, V., Pisarchik, A., and Hramov, A. (in press). Dynamics of map-based neuronal network with modified spike-timing-dependent plasticity, *Eur. Phys. J. Spec. Top.*

Aoki, S., Koyama, S., and Saito, T. (2018). Analysis and implementation of simple dynamic binary neural networks, *Proc. IJCNN*, pp. 471–476.

Apostolos, A., Dimitris, S., and Larger, L. (2005). Chaos-based communications at high bit rates using commercial fiber-optic links, *Nature* **438**, pp. 343– 346.

Appeltant, L., Soriano, M. C., Van der Sande, G., Danckaert, J., Massar, S., Dambre, J., Schrauwen, B., Mirasso, C. R., and Fischer, I. (2011).

Information processing using a single dynamical node as complex system, *Nat. Commun.* **2**, p. 468.

Arena, P., Baglio, S., Fortuna, L., and Manganaro, G. (1995). Hyperchaos from cellular neural networks, *Electronics Letters* **31**, 4, pp. 250–251.

Bai, C., Ren, H. P., Grebogi, C., and Baptista, M. (2018). Chaos-based underwater communication with arbitrary transducers and bandwidth, *Appl. Sci.* **8**, p. 162.

Banks, J., Brooks, J., Cairns, G., Davis, G., and Stacy, P. (1992). On Devaney's definition of chaos, *Amer. Math. Monthly* **99**, pp. 332–334.

Bao, B., Bao, H., Wang, N., Chen, M., and Xu, Q. (2017). Hidden extreme multistability in memristive hyperchaotic system, *Chaos, Solitons and Fractals* **94**, pp. 102–111.

Barrett, T., Suzek, T., Troup, D., Wilhite, S., Ngau, W.-C., Ledoux, P., Rudnev, D., Lash, A., Fujibuchi, W., and Edgar, R. (2005). Ncbi geo: Mining millions of expression profiles - database and tools, *Nucleic Acids Research* **33**, DATABASE ISS., pp. D562–D566.

Barrio, R., Angeles Martínez, M., Serrano, S., and Shilnikov, A. (2014). Macro- and micro-chaotic structures in the Hindmarsh-Rose model of bursting neurons, *Chaos: An Interdisciplinary Journal of Nonlinear Science* **24**, 2, p. 023128.

Barrio, R. and Shilnikov, A. (2011). Parameter-sweeping techniques for temporal dynamics of neuronal systems: case study of Hindmarsh-Rose model, *IJ Mathematical Neuroscience* **1**.

Bastos, J. and Caiado, J. (2011). Recurrence quantification analysis of global stock, **390**.

BEA (2016). USA Recessions, Gross Domestic Product [A191RP1Q027SBEA] - US. Bureau of Economic Analysis, `https://fred.stlouisfed.org/series/A191RP1Q027SBEA`, retrieved from FRED, Federal Reserve Bank of St. Louis; November 10, 2016.

Blakely, J. N., Hahs, D. W., and Corron, N. J. (2013). Communication waveform properties of an exact folded-band chaotic oscillator, *Physica D-Nonlinear Phenomena* **263**, 15 November, pp. 99–106, doi:10.1016/j.physd.2013.08.009.

Boccaletti, S., Pisarchik, A., del Genio, C., and Amann, A. (2018a). *Synchronization: From Coupled Systems to Complex Networks* (Cambridge Univ. Press).

Boccaletti, S., Pisarchik, A., del Genio, C., and Amann, A. (2018b). *Synchronization: From Coupled Systems to Complex Networks* (Cambridge University Press, Cambridge).

Bollt, E. and Santitissadeekorn, N. (2013). *Applied and Computational Measurable Dynamics* (SIAM).

Bose, B. K. (2007). Neural network applications in power electronics and motor drives - an introduction and perspective, *IEEE Trans. Ind. Electron.* **54**, pp. 14–33.

Brunton, S. L., Brunton, B. W., Proctor, J. L., Kaiser, E., and Kutz, J. N. (2017). Chaos as an intermittently forced linear system, *Nature Communications* **8**, p. 19.

Bucolo, M., Guo, J., Intaglietta, M., and Coltro, W. (2017). Guest editorial, special issue on microfluidics engineering for point-of-care diagnostics, *IEEE Trans. on Biomedical Circuits and Systems* **11**, pp. 1488–1499.

Budisic, M., Mohr, R., and Mezic, I. (2012). Applied koopmanism, *Chaos* **22**, p. 047510.

Buono, P.-L., Chan, B., Ferreira, J., Palacios, A., Reeves, S., Longhini, P., and In, V. (2018a). Symmetry-breaking bifurcations and patterns of oscillations in rings of crystal oscillators, *SIAM J. Appl. Dyn. Syst.* **17**, 2, pp. 1310–1352.

Buono, P.-L., In, V., Longhini, P., Olender, L., Palacios, A., and Reeves, S. (2018b). Phase drift on networks of coupled of crystal oscillators for precision timing, *Phys. Rev. E* **98**, p. 012203.

Burns, E. M. (2009). Long-term stability of spontaneous otoacoustic emissions, *J. Acoust. Soc. Am.* **125**, pp. 3166–3176.

Burns, S.-E. A. T. A., E. M. and Jones, K. (1984). Interactions among spontaneous otoacoustic emissions. i. distortion products and linked emissions, *Hearing Res.* **16**, pp. 271–278.

Cairone, F., Gagliano, S., and Bucolo, M. (2016a). Experimental study on the slug flow in a serpentine microchannel, *International Journal Experimental Thermal and Fluid Science* **76**, pp. 34–44.

Cairone, F., Gagliano, S., Carbone, D. C., Recca, G., and Bucolo, M. (2016b). Micro-optofluidic switch realized by 3d printing technology, *Microfluidics and Nanofluidics* **20:61**, pp. 1–10.

Cairone, F., Mirabella, D., Cabrales, P., Intaglietta, M., and Bucolo, M. (2018). Quantitative analysis of spatial irregularities in rbcs flows, *Chaos Solitons and Fractals* **115**, pp. 349–355.

Cairone, F., Ortiz, D., Cabrales, P., Intaglietta, M., and Bucolo, M. (2017a). Emergent behaviors in rbcs flows in micro-channels using digital particle image velocimetry, *Microvascular Research* **116**, pp. 77–86.

Cairone, F., Sanalitro, D., Ortiz, D., Cabrales, P., Intaglietta, M., and Bucolo, M. (2017b). Dpiv analysis of rbcs flows in serpentine micro-channel, in *2017 European Conference on Circuit Theory and Design (ECCTD)* (Catania, Italy).

Camalet, D. T.-J. F., S. and Prost, J. (2000). Auditory sensitivity provided by self-tuned critical oscillations of hair cells, *Proc. Natl. Acad. Sci. U.S.A.* **97**, pp. 3183–3188.

Campbell, S., Wang, D., and Jayaprakash, C. (1999). Synchrony and desynchrony in integrate-and-fire oscillators, *Neural Comput* **11**, pp. 1595–1619.

Campos-Cantón, E., Barajas-Ramírez, J. G., Solis-Perales, G., and Femat, R. (2010). Multiscroll attractors by switching systems, *Chaos: An Interdisciplinary Journal of Nonlinear Science* **20**, 1, p. 013116.

Carroll, T. L. (2017). Communication with unstable basis functions, *Chaos Solitons and Fractals* **104**, pp. 766–771, doi:10.1016/j.chaos.2017.09.039.

Channell, P., Fuwape, I., Neiman, A., and Shilnikov, A. (2009). Variability of bursting patterns in a neuronal model in the presence of noise, *IJ. Computational Neuroscience* **27**.

Chen, G. R. and Ueta, T. (1999). Yet another chaotic attractor, *Int. J. Bifurcation and Chaos* **9**, pp. 1465–1466.

Chen, W.-S. (2011). Use of recurrence plot and recurrence quantification analysis in taiwan unemployment rate time series, *Physica A: Statistical Mechanics and its Applications* **390**, 7, pp. 1332–1342.

Chua, L. (1971). Memristor-the missing circuit element, *IEEE Trans. Circuit Theory* **18**, pp. 507–519.

Chua, L. (2011). Resistance switching memoties are memristors, *Applied Physics A* **102**, pp. 765–783.

Chua, L. and Kang, S. (1976). Memristive devices and systems, *Proceedings of the IEEE* **375**, pp. 209–223.

Chua, L. O. (1992). *The Genesis of Chua's circuit* (Electronics Research Laboratory, College of Engineering, University of California Berkeley).

Clark, K.-D.-Z. P., W.W. and Bohne, B. (1984). Spontaneous otoacoustic emissions in chinchilla ear canals: correlation with histopathology and suppression by external tones, *Hearing Res.* **16**, pp. 299–314.

Cohen, S. and Gauthier, D. (2012). A pseudo-matched filter for chaos, *Chaos* **22**, p. 033148.

Cooper, N. and Rhode, W. (1997). Mechanical responses to two-tone distortion products in the apical and basal turns of the mammalian cochlea, *J. Neurophysiol.* **78**, pp. 261–270.

Cooper, S. (2004). Is whole-culture synchronization biology's 'perpetual-motion machine'? *Trends Biotechnol.* **22**, pp. 266–269.

Corron, N. and Blakely, J. (2015). Chaos in optimal communication waveforms, *Proc. R. Soc. A* **471**, p. 2180.

Corron, N. J., Blakely, J. N., and Stahl, M. T. (2010). A matched filter for chaos, *Chaos* **20**, 2, p. 023123.

Costamagna, E. and DiGialleonardo, E. (2018a). Mimiking railway vehicles hunting behaviors by means of sequence generators based on optimized sums of chaotic standard equations, *26th Conference on Nonlinear Dynamics of Electronic Systems NDES2018*.

Costamagna, E. and DiGialleonardo, E. (2018b). Progress in modeling railway hunting behaviors by means of chaotic equations, *IEEE Conferences, 2018 Workshop on Complexity in Engineering (COMPENG)*.

Costamagna, E., Favalli, L., and Gamba, P. (2005). Multipath channel modeling with chaotic attractors, *Proc. IEEE* **90**, pp. 842–859.

Costamagna, E., Favalli, L., and Tarantola, F. (2003). Modeling and analysis of aggregate and single stream internet traffic, *IEEE Globecom 2003, IEEE Conferences* **7**, pp. 3830–3834.

Crowley, M. F. and Epstein, I. R. (1989). Experimental and theoretical studies of a coupled chemical oscillator: phase death, multistability and in-phase and out-of-phase entrainment, *The Journal of Physical Chemistry* **93**, 6, pp. 2496–2502.

Crowley, P. M. (2008). Analyzing convergence and synchronicity of business and growth cycles in the euro area using cross recurrence plots, *The European Physical Journal Special Topics* **164**, 1, pp. 67–84.

D'Addabbo, A., Palmieri, O., Latiano, A., Annese, V., Mukherjee, S., and Ancona, N. (2011). Rs-snp: a random-set method for genome-wide association studies, *BMC Genomics* **12**, 1, p. 166, doi:10.1186/1471-2164-12-166.

de Boer, E. (1976). On the 'residue' and auditory pitch perception, in *Auditory System Vol. 3 (Handbook of Sensory Physiology)* (Springer), pp. 479–583.

Derighetti, B., Ravani, M., Stoop, R., Meier, P., Brun, E., and Badii, R. (1985). Period-doubling lasers as small-signal detectors, *Phys. Rev. Lett.* **55**, pp. 1746–1748.

Devaney, R. L. (1989). *Introduction to Chaotic Dynamical Systems* (Addison-Wesley).

DiGialleonardo, E., Bruni, S., and True, H. (2014). Analysis of the nonlinear dynamics of a 2axle freight wagon in curves, *Vehicle System Dynamics* **52**, pp. 125–141.

Digonnet, M. (2001). *Rare-Earth-Doped Fiber Lasers and Amplifiers* (Stanford University, Stanford, California).

Driscoll, T., Quinn, J., Klein, S., Kim, H., Kim, B., Pershin, Y., Ventra, M. D., and Basov, D. (2010). Memristive adaptive filters, *Applied Physics Letters* **97**, p. 093502.

Dudkowski, D., Jafari, S., Kapitaniak, T., Kuznetsov, N., Leonov, G., and Prasad, A. (2016). Hidden attractors in dynamical systems, *Phys. Rep.* **637**, pp. 1–50.

Echenausía-Monroy, J. L., García-López, J. H., Jaimes-Reategui, R., López-Mancilla, D., and Huerta-Cuellar, G. (2018). Family of bistable attractors contained in an unstable dissipative switching system associated to a snlf, *Complexity* **2018**.

Eckmann, J.-P., Kamphorst, S. O., and Ruelle, D. (1987). Recurrence plots of dynamical systems, *EPL (Europhysics Letters)* **4**, 9, p. 973.

Eguiluz, V., Ospeck, M., Choe, Y., Hudspeth, A., and Magnasco, M. (2000). Essential nonlinearities in hearing, *Phys. Rev. Lett.* **84**, pp. 5232–5235.

Fabretti, A. and Ausloos, M. (2005). Recurrence plot and recurrence quantification analysis techniques for detecting a critical regime. Examples from financial market inidices, *International Journal of Modern Physics C* **16**, 05, pp. 671–706.

Fastl, H. and Zwicker, E. (2007). *Psychoacoustics - Facts and Models* (Springer-Verlag, Berlin).

FIPS (2015a). Hmacsha-256: Hash based message authentication code, http://csrc.nist.gov/publications/fips/fips198-1/FIPS-198-1_final.pdf, accessed: 2018-10-26.

FIPS (2015b). Sha-256: Secure hash algorithm, https://nvlpubs.nist.gov/nistpubs/FIPS/NIST.FIPS.180-4.pdf, accessed: 2018-10-26.

FitzHugh, R. (1961). Impulses and physiological states in theoretical models of nerve membrane, *Biophys. J.* **1**, pp. 445–466.

Fortuna, L., Frasca, M., and Xibilia, M. G. (2009). *Chua's Circuit Implementation: Yesterday, Today and Tomorrow* (World Scientific, Singapore).

Fruth, F., Jülicher, F., and Lindner, B. (2014). An active oscillator model describes the statistics of spontaneous otoacoustic emissions, *Biophys. J.* **107**, pp. 815–824.

Gardiner, C. W. (2003). *Handbook of Stachastic Methods 3rd Ed.* (Springer: Complexity).

Gerasimova, S., Gelikonov, G., Pisarchik, A., and Kazantsev, V. (2015). Synchronization of optically coupled neural-like oscillators, *J. Commun. Technol. Electron.* **60**, pp. 900–903.

Gilardi-Velázquez, H. E., Ontañón-García, L. J., Hurtado-Rodriguez, D. G., and Campos-Cantón, E. (2017). Multistability in piecewise linear systems versus eigenspectra variation and round function, *International Journal of Bifurcation and Chaos* **27**, 09, p. 1730031.

Gold, T. (1948). Hearing. ii. the physical basis of the action of the cochlea, *Proc. R. Soc. Lond. B Biol. Sci.* **135**, pp. 492–498.

Gomez, F., Lorimer, T., and Stoop, R. (2016). Signal-coupled subthreshold Hopf-type systems show a sharpened collective response, *Phys. Rev. Lett.* **116**, p. 108101.

Gomez, F., Saase, V., Buchheim, N., and Stoop, R. (2014). How the ear tunes in to sounds: A physics approach, *Phys. Rev. Appl.* **1**, p. 014003.

Gomez, F. and Stoop, R. (2014). Mammalian pitch sensation shaped by the cochlear fluid, *Nat. Phys.* **10**, pp. 530–536.

Gorban, A. N., Smirnova, E. V., and Tyukina, T. A. (2010). Correlations, risk and crisis: From physiology to finance, *Physica A: Statistical Mechanics and its Applications* **389**, 16, pp. 3193–3217.

Gray, C., König, P., Engel, A., and Singer, W. (2013). Oscillatory responses in cat visual cortex exhibit inter-columnar synchronization which reflects global stimulus properties, *Nature* **338**, pp. 334–337.

Gray, D. and Michel, A. N. (1992). A training algorithm for binary feed forward neural networks, *IEEE Trans. Neural Netw.* **3**, pp. 176–194.

Grosse-Erdmann, K.-G. and Manguillot, A. P. (2011). *Linear Chaos* (Springer).

Hagan, M., Demuth, H., and Beale, M. (1996). *Neural Network Design*, 3rd edn. (Pws Pubs.).

Hanakawa, T., Dimyan, M. A., and Hallett, M. (2008). Motor planning, imagery, and execution in the distributed motor network: a time-course study with functional mri, *Cereb. Cort.* **18**, 12, pp. 2775–2788.

Haykin, S. (1994). *Communication Systems*, 3rd edn. (Wiley, New York).

Haykin, S. S. (2009). *Neural Networks and Learning Machines*, Vol. 3 (Pearson Upper Saddle River, NJ).

Hindmarsh, J. and Rose, R. (1984). A model of neuron bursting using three first order differential equations, *Proc. R. Soc. Lond. B.* **221**, pp. 87–102.

Hodgkin, A. and Huxley, A. (1952). A quantitative description of membrane current and its application to conduction and excitation in nerve, *J. Physiol.* **117**, pp. 500–544.

Hoffman, M. (2008). On the dynamics of european two-axle railway freight wagons, *Nonlinear Dynamics* **52**, pp. 301–311.

Holling, C. S. (2001). Understanding the complexity of economic, ecological, and social systems, *Ecosystems* **4**, 5, pp. 390–405.

Huerta-Cuellar, G., Jimenez-Lopez, E., Campos-Cantón, E., and Pisarchik, A. N. (2014). An approach to generate deterministic brownian motion, *Commu-*

nications in Nonlinear Science and Numerical Simulation **19**, 8, pp. 2740–2746.

Ignatiadis, K. (2018). *Dissonance Perception in the Hopf Cochlea* (PhD thesis, ETH Zürich).

Institute, T. S. (2014). http://www.turkstat.gov.tr/Start.do.

Itoh, M. and Chua, L. (2008). Memristor oscillators, *Int J Bifurc Chaos* **18**, pp. 3183–3206.

Izhikevich, E. (2006). *Dynamical Systems in Neuroscience* (MIT Press).

Izumi, S., Findley, T. W., Ikai, T., Andrews, J., Daum, M., and Chino, N. (1995). Facilitatory effect of thinking about movement on motor-evoked potentials to transcranial magnetic stimulation of the brain, *Am. J. Phys. Med. Rehab.* **74**, 3, pp. 207–213.

Jeannerod, M. (1994). The representing brain: neural correlates of motor intention and imagery, *Behav. Brain Sci.* **17**, pp. 187–202.

Jensen, C. N. and True, H. (1997). On a new route to chaos in railway dynamics, *Nonlinear Dynamics* **13**, pp. 117–129.

Jiang, X., Gripon, V., Berrou, C., and Rabbat, M. (2016). Storing sequences in binary tournament-based neural networks, *IEEE Trans. Neural Netw.* **27**, pp. 913–925.

Ju, H., Neiman, A., and Shilnikov, A. (2018). Bottom-up approach to torus bifurcation in neuron models, *arXiv preprint arXiv:1805.11719*.

Kaddoum, G. (2016). Wireless chaos-based communication systems: A comprehensive survey, *IEEE Access* **4**, pp. 2621–2648, doi:10.1109/ACCESS.2016.2572730.

Kaddoum, G. (2017). Wireless chaos-based communication systems: A comprehensive survey, *IEEE Access* **4**, pp. 2621–2648.

Kanders, K., Lorimer, T., Gomez, F., and Stoop, R. (2017). Frequency sensitivity in mammalian hearing from a fundamental nonlinear physics model of the inner ear, *Sci. Rep.* **7**, p. 9931.

Kasai, T., Kawai, S., Kawanishi, M., and Yahagi, S. (1997). Evidence for facilitation of motor evoked potentials (meps) induced by motor imagery, *Brain Research* **744**, pp. 147–150.

Kawashima, K., Fujiwara, K., Yamamoto, T., Sigeta, M., and Kobayashi, K. (1991). Electro-optical bistability and multistability in gaas/alas superlattices with different miniband widths, *Japanese Journal of Applied Physics* **30**, 9A, p. L1542.

Kemp, D. (1979). Evidence of mechanical nonlinearity and frequency selective wave amplification in the cochlea, *Arch. Oto-Rhino-Laryngol.*, pp. 37–45.

Kern, A. (2003). *A nonlinear biomorphic Hopf-amplifier model of the cochlea* (MSc thesis, University of Zürich and ETH Zürich).

Kern, A. and Stoop, R. (2003). Essential role of couplings between hearing nonlinearities, *Phys. Rev. Lett.* **91**, p. 128101.

Kolumbán, G. (1998). Performance evaluation of chaotic communications systems: Determination of low-pass equivalent model, in *Proc. Nonlinear Dynamics of Electronic Systems (NDES'98)* (Budapest, Hungary), pp. 41–51.

Kolumbán, G. (2018). Concept of software defined electronics (SDE): A revolutionary new approach for researching, building and teaching of ICT systems, PM8 half-day tutorial, *IEEE International Symposium on Circuits and Systems (ISCAS'18)* (Florence, Italy), https://drive.google.com/drive/folders/1-Q-3OPZtqACy-JBsVhKqcB-UcOugDAKe?usp=sharing.

Kolumbán, G., Krébesz, T., and Lau, F. C. M. (2012). Theory and application of software defined electronics: Design concepts for the next generation of telecommunications and measurement systems, *IEEE Circuits and Systems Magazine* **12**, 2, pp. 8–34.

Koopman, B. O. (1931). Hamiltonian systems and transformations in Hilbert space, *Proc. Natl. Acad. Sci. U.S.A.* **17**, p. 315.

Koyama, S., Aoki, S., and Saito, T. (2018). Simple feature quantities for analysis of periodic orbits in dynamic binary neural networks, *IEICE Trans. Fund.* **4**, pp. 727–730.

Kuznetsov, N., Leonov, G., and Vagaitsev, V. (2010). Analytical-numerical method for attractor localization of generalized chuas system, *IFAC Proceedings Volumes* **4**, 1.

Lacot, E., Stoeckel, F., and Chenevier, M. (1994). Dynamics of an erbium-doped fiber laser, *Phys. Rev. A* **49**, pp. 3997–4008.

Lebon, F., Byblow, W. D., Collet, C., Guillot, A., and Stinear, C. M. (2012). The modulation of motor cortex excitability during motor imagery depends on imagery quality, *Eur. J. Neurosci.* **35**, pp. 323–331.

Leonov, G. A. and Kuznetsov, N. V. (2013). Hidden attractors in dynamical systems: From hidden oscillation in Hilbert-Kolmogorov, Aizerman and Kalman problems to hidden chaotic attractor in Chua circuits, *Int. J. Bifurcation and Chaos* **23**, p. 1330002.

Levine, W. (2000). *The Control Handbook*, 2nd edn. (CRC Press).

Liang, N., Ni, Z., Takahashi, M., Murakami, T., Yahagi, S., Funase, K., Kato, T., and Kasai, T. (2007). Effects of motor imagery are dependent on motor strategies, *Neuroreport.* **18**, 12, pp. 1241–1245.

Liu, Z., Li, B., and Lai, Y.-C. (2012). Enhancing mammalian hearing by a balancing between spontaneous otoacoustic emissions and spatial coupling, *Europhys. Lett.* **98**, p. 20005.

Lonsbury-Martin, B., Martin, G., Probst, R., and Coats, A. (1988). Spontaneous otoacoustic emissions in a nonhuman primate. ii. cochlear anatomy, *Hearing Res.* **33**, pp. 69–93.

Lorenz, E. (1963). Deterministic nonperiodic flow, *Journal of the Atmospheric Sciences* **20**, pp. 130–141.

Lorimer, T., Gomez, F., and Stoop, R. (2015). Mammalian cochlea as a physics guided evolution-optimized hearing sensor, *Sci. Rep.* **5**, p. 12492.

Lozano, A., Rodriguez, M., and Roberto Barrio, R. (2016). Control strategies of 3-cell central pattern generator via global stimuli, *Sci. Rep.* **6**, p. 23622.

Luo, L. and Chu, P. (1998). Optical secure communications with chaotic erbium-doped fiber lasers, *J. Opt. Soc. Am. B* **15**, pp. 2524–2530.

Luo, L., Tee, T., and Chu, P. (1998a). Bistability of erbium-doped fiber laser, *Opt. Commun.* **146**, pp. 151–157.

Luo, L., Tee, T., and Chu, P. (1998b). Chaotic behavior in erbium-doped fiber-ring lasers, *J. Opt. Soc. Am. B* **15**, pp. 972–978.

Martignoli, S., Gomez, F., and Stoop, R. (2013). Pitch sensation involves stochastic resonance, *Sci. Rep.* **3**, p. 2676.

Martignoli, S. and Stoop, R. (2010). Local cochlear correlations of perceived pitch, *Phys. Rev. Lett.* **105**, p. 048101.

Martignoli, S., van der Vyver, J.-J., Kern, A., Uwate, Y., and Stoop, R. (2007). Analog electronic cochlea with mammalian hearing characteristics, *Appl. Phys. Lett.* **91**, p. 064108.

Meister, M., Wong, R., Baylor, D., and Shatz, C. (1991). Synchronous bursts of action potentials in ganglion cells of the developing mammalian retina, *Science* **252**, pp. 939–943.

Mets, A. D. (2014). Collaborative city co-design platform, `https://itea3.org/project/c3po.html`.

Milanović, V. and Zaghloul, M. (1996). Synchronization of chaotic neural networks and applications to communications, *Int. J. Bifurcat. Chaos* **6**, pp. 2571–2585.

Mitola, J. (1992). Software radio: Survey, critical analysis and future directions, in *Proc. IEEE National Telesystems Conference (NTC'92)* (Washington, DC), pp. 13/15–13/23.

Mizuguchi, N., Nakata, H., Uchida, Y., and Kanosue, K. (2012). Motor imagery and sport performance, *J. Phys. Fitness Sports Med.* **1**, 1, pp. 103–111.

Molaie, M., Jafari, S., Sprott, J. C., and Golpayegani, S. M. R. H. (2013). Simple chaotic flows with one stable equilibrium, *Int. J. Bifurcation and Chaos* **23**, p. 1350188.

Moloney, K. and Raghavendra, S. (2012). A linear and nonlinear review of the arbitrage-free parity theory for the cds and bond markets, in *Topics in Numerical Methods for Finance* (Springer), pp. 177–200.

National Bureau of Economic Research (2008). The NBER's recession dating procedure business cycle dating committee, `http://www.nber.org/cycles/jan2003.html`.

Neiman, A., Dierkes, K., Lindner, B., and Shilnikov, A. (2011). Spontaneous voltage oscillations and response dynamics of a Hodgkin-Huxley type model of sensory hair cells, *IJ. Mathematical Neuroscience* **1**.

Nickerson, K. W. (1973). Biological functions of multistable proteins, *Journal of Theoretical Biology* **40**, 3, pp. 507–515.

Nickles, D., Chen, H., Li, M., Khankhanian, P., Madireddy, L., Caillier, S., Santaniello, A., Cree, B., Pelletier, D., Hauser, S., Oksenberg, J., and Baranzini, S. (2013). Blood rna profiling in a large cohort of multiple sclerosis patients and healthy controls, *Human Molecular Genetics* **22**, 20, pp. 4194–4205.

Norton, S., Mott, J., and Champlin, C. (1989). Behavior of spontaneous otoacoustic emissions following intense ipsilateral acoustic stimulation, *Hearing Res.* **38**, pp. 243–258.

Ohyama, K., Wada, H., Kobayashi, T., and Takasaka, T. (1991). Spontaneous otoacoustic emissions in the guinea pig, *Hearing Res.* **56**, pp. 111–121.

Ontañón-García, L. J. and Campos-Cantón, E. (2017). Widening of the basins of attraction of a multistable switching dynamical system with the location of symmetric equilibria, *Nonlinear Analysis: Hybrid Systems* **26**, pp. 38–47.

Orlando, G. (2016). A discrete mathematical model for chaotic dynamics in economics: Kaldor's model on business cycle, *Mathematics and Computers in Simulation* **125**, pp. 83–98, doi:doi:10.1016/j.matcom.2016.01.001.

Orlando, G. (2018). Chaotic Business Cycles within a Kaldor-Kalecki Framework, *Nonlinear Dynamical Systems with Self-Excited and Hidden Attractors* doi:10.1007/978-3-319-71243-7_6, http://dx.doi.org/10.1007/978-3-319-71243-7_6.

Orlando, G. and Zimatore, G. (2017). Rqa Correlations on Real Business Cycles Time Series, *Proceedings of the Conference on Perspectives in Nonlinear Dynamics - 2016* **1**, pp. 35–41, doi:10.29195/iascs.01.01.0009, http://www.ias.ac.in/describe/article/conf/001/01/0035-0041.

Orlando, G. and Zimatore, G. (2018). Recurrence quantification analysis of business cycles, **110**, pp. 82–94, doi:10.1016/j.chaos.2018.02.032.

Ott, E. (1993). *Chaos in Dynamical Systems* (Cambridge University Press).

Papoulis, A. (1991). *Probability, Random Variables, and Stochastic Processes* (McGraw-Hill).

Parker, T. S. and Chua, L. O. (1987). Chaos: A tutorial for engineers, *Proceedings of the IEEE* **75**, pp. 982–1008.

Pecora, L. M. and Carroll, T. L. (1990). Synchronization in chaotic systems, *Physical Review Letters* **64**, Feb., pp. 821–824, doi:10.1103/PhysRevLett.64.821.

Perlman, R., Kaufman, C., and Speciner, M. (2016). *Network Security: Private Communication in a Public World* (Pearson Education India).

Pisarchik, A., Barmenkov, Y., and Kiryanov, A. (2003a). Experimental characterization of the bifurcation structure in an erbium-doped fiber laser with pump modulation, *IEEE J. Quantum Electron.* **39**, pp. 1567–1571.

Pisarchik, A., Barmenkov, Y., and Kiryanov, A. (2003b). Experimental demonstration of attractor annihilation in a multistable fiber laser, *Phys. Rev. E* **68**, p. 066211.

Pisarchik, A., Bashkirtseva, I., and Ryashko, L. (2017). Chaos can imply periodicity in coupled oscillators, *Eur. Phys. Lett.* **117**, p. 40005.

Pisarchik, A., Jaimes-Reátegui, R., and García-Vellisca, M. (2018). Asymmetry in electrical coupling between neurons alters multistable firing behavior, *Chaos* **28**, p. 033605.

Pisarchik, A., Kiryanov, A., Barmenkov, Y., and Jaimes-Reátegui, R. (2005). Dynamics of an erbium-doped fiber laser with pump modulation: theory and experiment, *J. Opt. Soc. Am. B* **22**, pp. 2107–2114.

Pisarchik, A. N. and Feudel, U. (2014). Control of multistability, *Physics Reports* **540**, 4, pp. 167–218.

Piskun, O. and Piskun, S. (2011). Recurrence quantification analysis of financial market crashes and crises, *arXiv preprint arXiv:1107.5420*.

Pusuluri, K., Pikovsky, A., and Shilnikov, A. (2017). Unraveling the chaos-land and its organization in the rabinovich system, in *Advances in Dynamics, Patterns, Cognition* (Springer), pp. 41–60.

Pusuluri, K. and Shilnikov, A. L. (2018). Homoclinic chaos and its organization in a nonlinear optics model, *arXiv preprint arXiv:1806.01309*.

Rasero, J., Cortes, J. M., Marinazzo, D., and Stramaglia, S. (2018). Connectome preprocessing by consensus clustering increases separability in group neuroimaging studies, *In press in Network Neuroscience*.

Rasero, J., Pellicoro, M., Angelini, L., Cortes, J. M., Marinazzo, D., and Stramaglia, S. (2017). Consensus clustering approach to group brain connectivity matrices, *Network Neuroscience* **1**, pp. 1–12.

Reategui, R. (2005). *Dynamic of Complex Systems with Parametric Modulation: Duffing Oscillators and a Fiber Laser* (Centro de Investigaciones en Optica, Leon, Gto., Mexico).

Reichenbach, T., Stefanovic, A., Nin, F., and Hudspeth, A. (2012). Waves on reissner's membrane: a mechanism for the propagation of otoacoustic emissions from the cochlea, *Cell Rep.* **1**, pp. 374–384.

Ren, H., Bai, C., Liu, J., Baptista, M., and Grebogi, C. (2016). Experimental validation of wireless communication with chaos, *Chaos* **26**, p. 083117.

Ren, H., Baptista, M., and Grebogi, C. (2013). Wireless communication with chaos, *Phys. Rev. Lett.* **110**, p. 184101.

Rhode, W. (1995). Interspike intervals as a correlate of periodicity pitch in cat cochlear nucleus, *J. Acoust. Soc. Am.* **97**, pp. 2414–2429.

Robles, L. and Ruggero, M. (2001). Mechanics of the mammalian cochlea, *Physiol. Rev.* **81**, pp. 1305–1352.

Robles, L., Ruggero, M., and Rich, N. (1997). Two-tone distortion on the basilar membrane of the chinchilla cochlea, *J. Neurophysiol.* **77**, pp. 2385–2399.

Romo-Aldana, J., García-López, J., Jaimes-Reategui, R., Sevilla, J., and Pisarchik, A. (2017). Automatización de equipo de laboratorio para el análisis del circuito electrónico del modelo neuronal de hindmarsh-rose, *SOMI XXXII* **4**, pp. 1–13.

Rössler, O. E. (1976). An equation for continuous chaos, *Phys. Lett. A* **57**, pp. 397–398.

Ruggero, M. (1992). Responses to sound of the basilar membrane of the mammalian cochlea, *Curr. Opin. Neurobiol.* **2**, pp. 449–456.

Ruggero, M., Rich, N., and Freyman, R. (1983). Spontaneous and impulsively evoked otoacoustic emission: indicators of cochlear pathology? *Hearing Res.* **10**, pp. 283–300.

Ruggero, M., Rich, N., Recio, A., Narayan, S., and Robles, L. (1997). Basilar membrane responses to tones at the base of the chinchilla cochlea, *J. Acoust. Soc. Am.* **101**, pp. 2151–2163.

Ruggero, M. and Temchin, A. (2002). The roles of the external, middle, and inner ears in determining the bandwidth of hearing, *Proc. Natl. Acad. Sci. USA* **99**, pp. 13206–13210.

Rulkov, N., Sushchik, M., Tsimring, L., and Volkovskii, A. (2001). Digital communication using chaotic-pulse-position modulation, *IEEE Trans. Circuits Systs. I* **48**, pp. 1436–1444.

Saito, T., Yamaoka, K., and Hamaguchi, T. (2017). Realization of desired digital spike-trains by a simple evolutionary algorithm, *IEICE* **4**, pp. 267–278.

Sanchez, F., LeFlohic, M., Stephan, G., LeBoudec, P., and Francois, P. (1995). Quasi-periodic route to chaos in erbium-doped fiber laser, *IEEE J. Quantum Electron.* **31**, pp. 481–488.

Sapuppo, F., Intaglietta, M., and Bucolo, M. (2008). Microfluidics real-time monitoring using cnn technology, *IEEE Transactions on Biomedical Circuits and Systems* **2**, pp. 78–87.

Sapuppo, F., Llobera, A., Schembri, F., Intaglietta, M., Cadarso, V. J., and Bucolo, M. (2010). A polymeric micro-optical interface for flow monitoring in biomicrofluidics, *Biomicrofluidics* **4**, pp. 1–13.

Sapuppo, F., Schembri, F., Fortuna, L., Llobera, A., and Bucolo, M. (2012). A polymeric micro-optical system for the spatial monitoring in two-phase microfluidics, *Microfluidics and Nanofluidics* **12**, pp. 165–174.

Sato, H., Sando, I., and Takahashi, H. (1991). Sexual dimorphism and development of the human cochlea: computer 3-d measurement, *Acta Oto-laryngol.* **111**, pp. 1037–1040.

Sato, R. and Saito, T. (2017). Stabilization of desired periodic orbits in dynamic binary neural networks, *Neurocomputing* **248**, pp. 19–27.

Saucedo-Solorio, J., Pisarchik, A., Kiryanov, A., and Aboites, V. (2003). Generalized multistability in a fiber laser with modulated losses, *J. Opt. Soc. Am. B* **20**, pp. 490–496.

Sausedo-Solorio, J. and Pisarchik, A. (2014). Synchronization of map-based neurons with memory and synaptic delay, *Phys. Lett. A* **378**, pp. 2108–2112.

Sausedo-Solorio, J. and Pisarchik, A. (2017). Synchronization in network motifs of delay-coupled map-based neurons, *Eur. Phys. J. Spec. Top.* **226**, pp. 1911–1920.

Service, T. S. M. (2018). Extreme maximum, minimum and average temperatures measured in long period, `https://mgm.gov.tr/eng/forecast-cities.aspx`.

Shabunin, A., Astakhov, V., Demidov, V., Provata, A., Baras, F., Nicolis, G., and Anishchenko, V. (2003). Modeling chemical reactions by forced limit-cycle oscillator: synchronization phenomena and transition to chaos, *Chaos Sol. Fract.* **15**, pp. 395–405.

Shera, C. (2003). Mammalian spontaneous otoacoustic emissions are amplitude-stabilized cochlear standing waves, *J. Acoust. Soc. Am.* **114**, pp. 244–262.

Shilnikov, A. (2012). Complete dynamical analysis of a neuron model, *Nonlinear Dynamics* **68**, 3, pp. 305–328.

Simonov, A., Gordleeva, S., Pisarchik, A., and Kazantsev, V. (2013). Synchronization with an arbitrary phase shift in a pair of synaptically coupled neural oscillators, *J. Exper. Theor. Phys. Lett.* **98**, pp. 632–637.

Sklar, B. (1988). *Digital Communications, Fundamentals and Applications* (Prentice Hall).

Smoorenburg, G. (1970). Pitch perception of two-frequency stimuli, *J. Acoust. Soc. Am.* **48**, pp. 924–942.

Solodkin, A., Hlustik, P., Chen, E. E., and Small, S. L. (2004). Fine modulation in network activation during motor execution and motor imagery, *Cereb. Cort.* **14**, pp. 1246–1255.

Sprott, J. C. (2010). *Elegant Chaos Algebraically Simple Chaotic Flows* (World Scientific, Singapore).

Stinear, C. M. and Byblow, W. D. (2003). Role of intracortical inhibition in selective hand muscle activation, *J. Neurophysiol.* **89**, pp. 2014–2020.

Stinson, D. (2005). *Cryptography: Theory and Practice*, 3rd edn. (Taylor & Francis Ltd.).

Stoop, R. and Gomez, F. (2016). Auditory power-law activation avalanches exhibit a fundamental computational ground state, *Phys. Rev. Lett.* **117**, p. 038102.

Stoop, R., Jasa, T., Uwate, Y., and Martignoli, S. (2007). From hearing to listening: Design and properties of an actively tunable electronic hearing sensor, *Sensors* **7**, pp. 3287–3298.

Stoop, R. and Kern, A. (2004). Two-tone suppression and combination tone generation as computations performed by the Hopf cochlea, *Phys. Rev. Lett.* **93**, p. 268103.

Strukov, D., Snider, G., Stewart, G., and Williams, R. (2008). The missing memristor found, *Nature* **453**, pp. 80–83.

Suykens, J. A. K. and Huang, A. (1997). A family of n-scroll attractors from a generalized chua's circuit, *Archiv fur Elektronik und Ubertragungstechnik* **51**, 3, pp. 131–137.

Suzuki, H., Imura, J., and Aihara, K. (2013). Chaotic ising-like dynamics in traffic signals, *Scientific Reports* **3**, p. 1127.

Talmadge, C., Long, G., Murphy, W., and Tubis, A. (1993). New off-line method for detecting spontaneous otoacoustic emissions in human subjects, *Hearing Res.* **71**, pp. 170–182.

Tang, W. K. S., Zhong, G. Q., Chen, G., and Man, K. F. (2001). Generation of n-scroll attractors via sine function, *IEEE Transactions on Circuits and Systems I: Fundamental Theory and Applications* **48**, 11, pp. 1369–1372.

Taulu, S. and Hari, R. (2009). Removal of magnetoencephalographic artifacts with temporal signal-space separation: Demonstration with single-trial auditory-evoked responses, *Human Brain Mapping* **30**, pp. 1524–1534.

Torikai, H., Funew, A., and Saito, T. (2008). Digital spiking neuron and its learning for approximation of various spike-trains, *Neural Networks* **21**, pp. 140–149.

Uchida, H. and Saito, T. (2017). Implementation of desired digital spike maps in the digital spiking neurons, in *ICONIP 2017, Part VI*, pp. 804–811.

Ulehlova, L., Voldrich, L., and Janisch, R. (1987). Correlative study of sensory cell density and cochlear length in humans, *Hearing Res.* **28**, pp. 149–151.

Veitch, D. and Abry, P. (1999). A wavelet-based joint estimator of the parameters of long-range dependence, *IEEE Trans. Inform. Theory* **45**, pp. 878–897.

Ventra, M., Pershin, Y., and Chua, L. (2009). Circuit elements with memory: memristors, memcapacitors, and meminductors, *Proceedings of the IEEE* **97**, pp. 1717–1724.

Vilfan, A. and Duke, T. (2008). Frequency clustering in spontaneous otoacoustic emissions from a lizard's ear, *Biophys. J.* **95**, pp. 4622–4630.

Volos, C., Kyprianidis, I., and Stouboulos, I. (2011). The memristor as an electric synapse synchronization phenomena, *Proc. Int. Conf. DSP2011 (Confu, Greece)* **1**, pp. 1–6.

Š., P. (2010). Integer rounding functions.

Wada, W., Kuroiwa, J., and Nara, S. (2002). Completely reproducible description of digital sound data with cellular automata, *Phys. Lett. A.* **306**, pp. 110–115.

Wang, G., Jiang, S., Wang, X., Shen, Y., and Yuan, F. (2016). A novel memcapacitor model and its application for generating chaos, *Mathematical Problems in Engineerings* **2016**, p. 3173696.

Wang, X. and Chen, G. (2012). A chaotic system with only one stable equilibrium, *Commun. Nonlinear Sci. Numer. Simulat.* **17**, pp. 1264–1272.

Wiesenfeld, K. and McNamara, B. (1985). Period-doubling systems as small-signal amplifiers, *Phys. Rev. Lett.* **55**, pp. 13–16.

Wiesenfeld, K. and McNamara, B. (1986). Small-signal amplification in bifurcating dynamical systems, *Phys. Rev. A* **33**, pp. 629–642.

Wio, S., Deza, R., and Lopez, M. (2012). *An Introduction to Stochastic Processes and Nonequilibrium Statistical Physics* (World Scientific Publishing).

Wu, A., Zhang, J., and Zeng, Z. (2011). Dynamical behaviors of a class of memristor-based Hopfield networks, *Physics Letters A* **375**, pp. 1661–1665.

Xu, Y., Wang, H., Li, Y., and Pei, B. (2014). Image encryption based on synchronization of fractional chaotic systems, *Commun. Nonlinear Sci.* **19**, pp. 3735–3744.

Yalcin, M. E., Suykens, J. A. K., and Vandewalle, J. (2000). Hyperchaotic n-scroll attractors, in *Nonlinear Dynamics of Electronic Systems* (World Scientific), pp. 25–28.

Yang, J., Strukov, D.B. and Stewart, D. (2013). The memristor as an electric synapse synchronization phenomena, *Memristive Devices for Computing* **8**, pp. 13–24.

Yao, J. L., Li, C., Ren, H., and Grebogi, C. (2017). Chaos-based wireless communication resisting multipath effects, *Phys. Rev. E* **96**, p. 032226.

Yao, J. L., Sun, Y., Ren, H., and Grebogi, C. (2018). Experimental wireless communication using chaotic baseband waveform, *IEEE Trans. Vhe. Technol.* to appear.

Yin, S., Liu, Y., and Ding, M. (2016). Amplitude of sensorimotor mu rhythm is correlated with bold from multiple brain regions: a simultaneous eeg-fmri study, *Front. Human Neurosci.* **10**, p. 364.

Zapala, M. A. and Schork, N. J. (2006). Multivariate regression analysis of distance matrices for testing associations between gene expression patterns and related variables, *Proceedings of the National Academy of Sciences* **103**, 51, pp. 19430–19435.

Zimatore, G., Fetoni, A. R., Paludetti, G., Cavagnaro, M., Podda, M. V., and Troiani, D. (2011). Post-processing analysis of transient-evoked otoacoustic emissions to detect 4 khz-notch hearing impairment–a pilot study, *Medical Science Monitor: International Medical Journal of Experimental and Clinical Research* **17**, 6, p. MT41.

Zimatore, G., Garilli, G., Poscolieri, M., Rafanelli, C., Gizzi, F. T., and Lazzari, M. (2017). The remarkable coherence between two italian far away recording stations points to a role of acoustic emissions from crustal rocks for earthquake analysis, *Chaos: An Interdisciplinary Journal of Nonlinear Science* **27**, 4, p. 043101.

Zoubir, A. M. and Iskander, D. R. (2007). Bootstrap methods and applications: A tutorial for the signal processing practitioner, *IEEE - Signal Processing Magazine* **24**, pp. 10–19.

Zwicker, E. and Heinz, W. (1995). Zur Häufigkeitsverteilung der menschlichen Hörschwelle, *Acustica* **5**, pp. 75–80.